IUPAB Biophysics Series
sponsored by
The International Union of Pure and Applied Biophysics

Biological effects of ultraviolet radiation

IUPAB Biophysics Series
sponsored by
The International Union of Pure and Applied Biophysics
Editors
Franklin Hutchinson
Yale University
Watson Fuller
University of Keele
Lorin J. Mullins
University of Maryland

Biological effects of ultraviolet radiation

WALTER HARM

Professor of Biology
University of Texas at Dallas

CAMBRIDGE UNIVERSITY PRESS
CAMBRIDGE
LONDON NEW YORK NEW ROCHELLE
MELBOURNE SYDNEY

Published by the Press Syndicate of the University of Cambridge
The Pitt Building, Trumpington Street, Cambridge CB2 1RP
32 East 57th Street, New York, NY 10022, USA
296 Beaconsfield Parade, Middle Park, Melbourne 3206, Australia

© Cambridge University Press 1980

First published 1980

Printed in the United States of America
Typeset by Automated Composition Service, Inc., Lancaster, Pennsylvania
Printed and bound by the Murray Printing Company, Westford, Massachusetts

Library of Congress Cataloging in Publication Data
Harm, Walter, 1925–
Biological effects of ultraviolet radiation.
(IUPAB biophysics series; 1)
1. Ultra-violet rays – Physiological effect.
2. Radiogenetics. I. Title. II. Series: Inter-
national Union for Pure and Applied Biophysics. IUPAB
biophysics series; 1.
QH652.H28   574.1'9154   77-88677
ISBN 0 521 22121 8 hard covers
ISBN 0 521 29362 6 paperback

CONTENTS

FOREWORD

The origins of this series were a number of discussions in the Education Committee and in the Council of the International Union of Pure and Applied Biophysics (IUPAB). The subject of the discussions was the writing of a textbook in biophysics; the driving force behind the talks was Professor Aharon Katchalsky, first while he was president of the Union, and later as the honorary vice-president.

As discussions progressed, the concept of a unified text was gradually replaced by that of a series of short inexpensive volumes, each devoted to a single topic. It was felt that this format would be more flexible and more suitable in light of the rapid advances in many areas of biophysics at present. Instructors can use the volumes in various combinations according to the needs of their courses; new volumes can be issued as new fields become important and as current texts become obsolete.

The International Union of Pure and Applied Biophysics was motivated to participate in the publication of such a series for two reasons. First, the Union is in a position to give advice on the need for texts in various areas. Second, and even more important, it can help in the search for authors who have both the specific scientific background and the breadth of vision needed to organize the knowledge in their fields in a useful and lasting way.

The texts are designed for students in the last years of the standard university curriculum and for Ph.D. and M.D. candidates taking advanced courses. They should also provide a suitable introduction for someone about to begin research in a particular field of biophysics. The Union is pleased to collaborate with the Cambridge University Press in making these texts available to students and scientists throughout the world.

Franklin Hutchinson, Yale University
Watson Fuller, University of Keele
Lorin Mullins, University of Maryland
Editors

PREFACE

The killing of cells by ultraviolet (or UV) radiation, in particular by wavelengths present in sunlight, has been studied for more than a century. The ultraviolet's mutagenic action was experimentally established not long after discovery of the mutagenicity of ionizing radiations by H. J. Muller in 1927. Concomitantly, UV action spectra for the killing of cells and viruses and for mutagenesis indicated that these effects were due to energy absorption in nucleic acids, substances then known to be part of the chromosomes, and as it turned out later, those actually carrying the hereditary information. Aside from such early roots, however, UV photobiology is a relatively recent area of research, comprising in essence the achievements of the past 20 to 30 years. Its development during this period paralleled the rapid progress in genetics and molecular biology, resulting mainly from the experimental use of microorganisms and viruses.

Basically, research in UV photobiology has employed the radiation as a tool to damage organisms in a fairly specific manner, at least more specific than ionizing radiations do. Investigating the consequences of such damage not only permitted conclusions regarding the UV-absorbing material and the ways in which the damage interferes with vital cellular processes, but also led to the observation of recovery effects. Work by A. Hollaender in the middle 1930s gave the first evidence for them, but their general significance was not recognized until the later forties and early fifties. The study of recovery phenomena indicated that, in contrast to earlier concepts, the eventual fate of an irradiated organism is highly conditional, rather than being fully determined at the time of energy absorption. Not only did these achievements offer explanations for otherwise irreconcilable discrepancies between the results obtained in different laboratories, subsequent studies on recovery phenomena also revealed their molecular basis: the existence of a variety of sophisticated and highly efficient repair processes enabling the cells to cope with otherwise unbearable conditions. They are of great importance in protecting organisms against radiation damage from our natural UV source, the sun, as well as against damage from a wide variety of other agents. Moreover, it is now evident that some of the reaction steps involved in repair play at the same time a significant role in general cellular maintenance processes and such basic phenomena as genetic recombination and DNA replication. The use of UV radiation has been basically involved in these important discoveries.

When Franklin Hutchinson, the chairman of the Editorial Board of this series of biophysics textbooks, approached me with regard to writing a book on *Biological Effects of Ultraviolet Radiation*, I accepted with mixed feelings. On the one hand, I was grateful for the opportunity to summarize the results, as well as my own views and thoughts, in an area of research that I have been associated with for the past 25 years, and in which my writing had so far been only in the form of original papers and review articles. On the other hand, it was obvious that the task of covering a field in which interpretations and concepts are still in a considerable flux, and to which hundreds of scientists contribute the results of their research almost daily, is a difficult one. My final decision was facilitated by the editor's request that the textbook be short, comprising essentially what he felt I carry around in my head. Although the latter turned out to be too optimistic in many instances, I was aware that the more time-consuming parts of the job would be: (1) the decision as to what one can consider of sufficient importance to provide graduate and advanced undergraduate students, or scientists established in other areas of research, with a useful background for entering this field; (2) the presentation of the material in a form easy to comprehend, particularly with regard to those potential readers unfamiliar with the inherent genetic and quantitative approaches. My apologies, if I have not always succeeded in these respects.

Among the rewards for writing a textbook is the author's satisfaction in presenting the facts, concepts, and his own thoughts in his field of competence in a didactically most desirable form. Full justification of the effort, however, requires the existence of a definitive need for such a book. The implicit criterion that the book must differ significantly from others on similar topics is not difficult to meet in the present case. There exist several volumes with contributions from research symposia and excellent review articles on various aspects of UV photobiology, published in journals, in periodicals, or in the form of multiauthored books. They are invaluable as a source of information for the advanced scientist but necessarily inadequate as introductory texts. A recently published work, *The Science of Photobiology*, edited by K. C. Smith (Plenum, 1977) covers the whole discipline of photobiology, including photosynthesis, photomorphogenesis, bioluminescence, and many other fields, and contains excellent contributions from many authors in their specific areas of competence. But only the chapter "Ultraviolet Radiation Effects on Molecules and Cells" coincides to a major extent with the contents of this book. Among textbooks, *Introduction to Research in Ultraviolet Photobiology* by J. Jagger (Prentice-Hall, 1967) puts its main emphasis on the techniques involved in UV-photobiological research. *Molecular Photobiology* by K. C. Smith and P. C. Hanawalt (Academic Press, 1969) covers a similar area, as does this book. However, almost 10 years have passed since publication of *Molecular Photobiology*, and in their contents the two books complement one another in several respects. Unquestionably, the emphasis on biological problems in the present book reflects the author's own research preference and his original training as a biologist.

The first three chapters of the book are designed to provide the reader with the background in chemistry and physics essential to an understanding of the biological effects, to which all of the remaining 10 chapters are devoted. The reader may feel that Chapters 7 and 8, dealing with recovery, repair processes, and related phenomena observed after UV damage, are treated in more detail than seems appropriate in comparison with other chapters. Perhaps this amounts to overemphasis of my own research area. But in all fairness, one can probably say that during the past 10 to 20 years hardly any other branch of UV photobiology contributed more to our general biological knowledge than the study of repair and recovery.

It is my pleasure to acknowledge the help of many colleagues, and the publishers of their work, for the kind permission to use their graphs or other illustrations in this book. The names are too numerous to mention them all here; acknowledgments are given at the appropriate places. Particular thanks are owed Dr. Franklin Hutchinson for the encouragement to my writing, and my colleague Dr. Claud S. Rupert, with whom I have had for the past 15 years close scientific relations in several areas covered by the book. These resulted in a number of joint publications as well as in an earlier attempt at writing together a comprehensive monograph on UV photobiology of microorganisms. We abandoned this effort, simply because the literature was turning out results faster than we were able to digest them for the purpose of writing, taking into account all other commitments and problems at a newly established university campus. Nevertheless, this apparently futile exercise was of considerable help in the preparation and writing of the present text, notably the first three chapters, which cannot deny the physicist's influence.

My thanks also include Dr. Michael H. Patrick for critically reviewing several chapters of the book, and Mr. H. Thomas Steely, Jr., a graduate student in the Molecular Biology Program at The University of Texas at Dallas for reading the manuscript and giving me his views as a potential user of the textbook. I am particularly indebted to Sally Rahn for secretarial help and for the typing of the manuscript.

Walter Harm

August 1979.

1 Introduction to ultraviolet radiation

1.1 General characteristics

Ultraviolet radiation or ultraviolet light (UV) is part of the spectrum of electromagnetic waves, covering the *interval between X-rays and visible light*. Although real boundaries between these different kinds of radiation do not
exist in physical terms, they are dictated by practical considerations. Percep-
tion by the human eye begins at about 380 nm,[1] which therefore constitutes
the upper wavelength limit of the UV spectrum. Specifying a lower wavelength limit is far more arbitrary, but a reasonable figure is 100 nm, below
which radiations ionize virtually all molecules. For the study of biological
effects of UV radiation there is a practical lower limit at about 190 nm.
Shorter wavelengths are strongly absorbed by water and air, making it mandatory to irradiate in vacuum (vacuum ultraviolet). Not only does this require
special equipment, it is also incompatible with experimentation on most
living systems. Therefore, with few exceptions, we consider for the biological
effects of UV radiation essentially the wavelength range from 190 to 380 nm.

Electromagnetic radiations transfer their energy in units of *energy quanta*,
or *photons*. As first stated by Planck,

$$E = h\nu \tag{1.1}$$

that is, the energy of a photon (E) is directly proportional to its frequency of
vibrations per second (ν), with h being Planck's constant (6.62×10^{-27} erg ·
sec or 6.62×10^{-34} Joule · sec). Because $\nu = v/\lambda$ (where v is the velocity of
light, and λ is the wavelength), equation (1.1) can also be written in the form

$$E = h v / \lambda \tag{1.2}$$

E and ν of a photon are independent of the medium transmitting the radiation, whereas v, and consequently λ, varies. It is nevertheless customary to
characterize UV radiation in terms of its wavelength, namely, by specifying
the wavelength that the radiation would have in vacuum, where $v = 2.99 \times
10^{10}$ cm/sec.

For expressing the energy of a single UV photon, the erg and the Joule
(J) are rather large units. Conventionally one uses the electronvolt (eV),
defined as the energy gained by an electron in passing through a potential
difference of 1 volt, which equals 1.6×10^{-12} erg or 1.6×10^{-19} J. Ac-

1

cording to equation (1.2), 1 eV corresponds to a photon of 1.24×10^{-4} cm (or 1240 nm) wavelength, and it follows from the inverse proportionality of the energy of a photon with its wavelength that

$$E[\text{eV}] = \frac{1240}{\lambda[\text{nm}]} \qquad (1.3)$$

For photochemical purposes it is sometimes useful to express photon energies in kilocalories per einstein because absorption of 1 einstein ($= 6.02 \times 10^{23}$ photons) can excite 1 mole of the absorbing substance. In these terms, the photon energy (E), and thus the excitation energy of a molecule (E_{exc}), are related to the wavelength by

$$E[\text{kcal/einstein}] = E_{\text{exc}}[\text{kcal/mole}] = \frac{28,590}{\lambda[\text{nm}]} \qquad (1.4)$$

From equations (1.3) and (1.4) it follows that the energies of photons within the 190 to 380 nm wavelength range vary by at most a factor of 2, namely, from 6.5 to 3.3 eV, or from 150 to 75 kcal/einstein.

Radiation of the ultraviolet and the adjacent visible spectral range (as well as all other less energetic radiation) is summarily called *nonionizing radiation*, as opposed to *ionizing radiation*. The latter is represented in the electromagnetic spectrum essentially by X-rays and gamma-rays; other kinds of ionizing radiations (such as beta-rays, alpha-rays, protons, etc.) consist of ionizing particles.

The main reason for this distinction is their interaction with matter: Ionizing radiations, in contrast to nonionizing radiations, are capable of ionizing all kinds of atoms and molecules. Absorption of nonionizing radiations typically leads to *electronic excitation* of atoms and molecules (see Section 1.2); however, ionization already begins in the ultraviolet spectral region around 200 nm and, depending on the type of atoms or molecules, becomes more relevant as the wavelengths further decrease. Most organic molecules require wavelengths below 150–180 nm in order to dissociate an electron (and thus to leave behind a positive ion). In view of the wavelengths used in most biological UV experiments and the molecules primarily affected by the energy absorption, we can for all practical purposes, exclude ionizations as one of the possible immediate consequences of UV-irradiation in biological materials.

1.2 Electronic excitation

An atom or molecule absorbing a UV photon assumes for a period of 10^{-10} to 10^{-8} sec an *excited state*, in which the energy of the electrons is increased by the amount of photon energy. Because the number of possible energy states for the electrons of an atom or molecule is finite, only photons of specific energies (i.e., specific wavelengths) can be absorbed by an isolated atomic or molecular species. Consequently, UV absorption spectra of gas atoms at low pressure usually consist of sharply defined, discrete *absorption lines*. In simple molecules, whose rotational and vibrational energy states can also be affected by the absorbed photon, the spectral lines occur in closely spaced groups, or *absorption bands*. The larger and the more complex the molecule, the more closely spaced become the line patterns of the bands. In the solid or liquid state, where interactions with neighboring molecules prevent free rotation and disturb the energy levels, the fine details disappear from the observed absorption spectrum, and the bands become a *continuum* of smoothly changing intensity with the wavelength. This is characteristic of the UV-absorption spectra of the most important biomolecules, such as nucleic acids and proteins (see Figure 3.2).

The excitation energy provided by UV photons is much higher than the energy of thermal motions of the molecules at physiological temperatures. The latter is of the order of Boltzmann's constant times the absolute temperature, which, at 27°C, amounts to only 0.026 eV/molecule (= 0.60 kcal/mole), in contrast to the 3.3 to 6.5 eV/molecule (or 75 to 150 kcal/mole) available from UV absorption. Consequently, the absorbing molecules temporarily assume energy levels that otherwise they would never attain and thus acquire properties differing considerably from those effective in ordinary chemistry.

The lifetime of a molecule in its usual excited state (10^{-10} to 10^{-8} sec), which is still long compared with the time required for the energy absorption itself (approximately 10^{-15} sec), can be greatly extended if the excited electron is trapped in an (energetically somewhat lower) *triplet* excited state. In contrast to the usual *singlet* state, the triplet state is characterized by two electrons with *unpaired spin*. Because the return from the triplet state to the ground state is "forbidden" (i.e., occurs at a low probability), the triplet may last 10^{-3} sec or even longer and is, therefore, called *metastable*.

As an excited electron returns to a lower energetic state, its excess energy may be disposed of in several ways:

1. It can be emitted as a photon, resulting in *fluorescence*. Fluorescent light is recognized by its usually longer wavelength, compared with the exciting radiation. Emission from molecules in the metastable excited state occurs over a longer period of time and is called *phosphorescence*.
2. The excitation energy can be *dissipated as thermal energy* in the course of collisions with other molecules.
3. The energy may cause the excited molecule to undergo a *photochemical reaction* that otherwise would not occur. The likelihood for this to happen increases with the lifetime of the excited state and is thus greatly enhanced for the triplet state. Photochemical reactions are the immediate effects of UV radiation in biologically relevant molecules, and constitute the basis for the observed photobiological phenomena. They will be discussed in more detail in Chapter 3.

1.3 Biological effectiveness

Although the photon energies within the biologically applicable UV spectrum vary by no more than a factor of 2, equal numbers of incident photons can cause photochemical (and consequently photobiological) reactions differing in quantity by several orders of magnitude. This rather general observation indicates wide variations in the absorption of photons at different wavelengths by the relevant biomolecules. Obviously only absorbed, but not transmitted or reflected, radiation energy can be photochemically effective (Draper-Grotthus principle). We will see later that the majority of biological UV effects are due to photochemical reactions in nucleic acids, which constitute the genetic material of all cellular organisms and viruses. Protein effects generally play a minor role, but are relevant in some cases.

Nucleic acids and most proteins have their absorption maxima well below 300 nm and absorb little at wavelengths above 300 nm. Therefore, it is not surprising that biological UV effects produced by radiation below 300 nm are infrequently observed at longer wavelengths. For this reason, it is customary to subdivide the UV spectrum into *near UV* (300–380 nm) and *far UV* (below 300 nm), the adjectives near and far indicating the relative distance from the visible spectral range. Because far UV is much more effective than near UV with respect to inactivation of microorganisms (which is one of the technical applications of UV radiation), it is also called *germicidal UV*.

To separate the far and near UV region just at 300 nm is convenient, but the borderline could as well be placed somewhat above or below this point. As a matter of fact, the region *surrounding* 300 nm (from approximately 290 to 315 nm) is in several respects very critical. In this region absorption of most nucleic acids and proteins diminishes rapidly, so that it becomes unmeasurable at still higher wavelengths. Conversely, the solar emission spectrum reaching the earth's surface contains mainly near UV (besides visible and infrared light); the shortest wavelengths are usually somewhere between 290 and 315 nm, depending on many factors (see Section 11.1). Furthermore, the short wavelength cutoff in the transmission of some of the commercial glasses falls into this region.

If, for simplicity, one wants to identify measurable nucleic acid absorption with the far-UV range, and to identify the solar emission spectrum with the near-UV range, one would have to say that the two ranges overlap in the small region from approximately 290 to 315 nm. Not only does this avoid semantic confusion, but it also emphasizes the fact that effects characteristic of both far and near UV irradiation may occur in this region at comparable rates. This overlap region is indeed of considerable theoretical and practical significance. The medical literature, primarily concerned with the dermatological effects of UV radiation, often distinguishes three UV spectral regions: UV-A (>315 nm), UV-B (280–315 nm), and UV-C (<280 nm), with most interaction between sunlight and the human skin occurring in the UV-B region. Although such a subdivision is certainly useful, it has not become popular in the biophysical literature.

1.4 Sources of ultraviolet radiation

For reasons pointed out previously, most biological experiments require a UV source emitting appreciably in the far UV. Thus the lamp envelopes and any optical components of the irradiation equipment must consist of *quartz* or material of similar transmittance because ordinary glasses are opaque for far-UV wavelengths. The following paragraphs will briefly summarize some technical facts that are fundamental for understanding the text and for carrying out simple experimental work in UV photobiology. For more detailed information in this regard the reader is referred to the textbook by Jagger (1967).

1.4.1 UV sources with broad spectral emission

Early experimental studies of biological UV effects frequently employed UV sources with a broad continuous emission spectrum, for example: *hydrogen-*, *xenon-*, or *high-pressure mercury-vapor* discharge lamps. There has been some virtue in the fact that with this type of equipment one is less likely to overlook effects characteristic of only a small region of the UV spectrum. Furthermore, a broad spectral range may sometimes be favored for solving problems of an applied nature. However, regarding basic UV-photobiological research, the present state of knowledge usually requires a quantitative correlation of the observed biological effects with defined, limited spectral regions.

1.4.2 Limitation of spectral regions

A broad emission spectrum can be narrowed by inserting appropriate *optical glass filters* or *liquids.* Suitable combinations of such filters may be chosen for transmission of a rather confined wavelength region, or for the selection of a single wavelength from a line emission spectrum.

1.4.3 Monochromatic UV radiation

Commercially available *monochromators* with quartz-transmission or aluminum-reflection optics are generally used in biological experiments for isolating a single wavelength from a line emission spectrum or a narrow wavelength band from a continuum. If such an instrument is not available, *interference filters*, preferably in combination with a UV source providing a suitable line emission spectrum, can be used to achieve adequate spectral monochromasy. Such filters are now also available for far-UV wavelengths (see Appendix in Jagger, 1967).

1.4.4 Germicidal lamps

Radiation at 254 nm (or, more accurately, 253.7 nm) is obtained at high intensity from low-pressure mercury-vapor discharge lamps. Because of their commercial use for sterilizing air, water, and so on, they are widely known as *germicidal lamps*. As shown in Table 1.1, approximately 95 percent of the total UV emission of a germicidal lamp, or more than 97 percent of the far-UV emission, is at 254 nm. Because in addition this wavelength is nearly maximally effective for many UV-photobiological experiments (owing to its closeness to the absorption maximum of nucleic acids), such a lamp can be considered, for most experimental purposes, a *quasi-monochromatic UV source*. Because of the low cost and ease of handling of these lamps, they are used for the majority of published work; in fact, they are in many laboratories the only source of UV radiation.

It is important to notice that some types of germicidal lamps produce *ozone* from oxygen in the air, which is easily smelled in proximity to the lamp. Ozone production is essentially due to emission at 185 nm, a mercury line that is transmitted through a lamp envelope consisting of pure quartz. Lamp envelopes made of Vycor, or of quartz with mineral impurities, do not transmit 185 nm; therefore, these lamps are called ozone-free. Because the biological effects of 185 nm UV are expected to differ substantially, both in quantity and quality, from those of 254 nm, and because the in-

Table 1.1. *Distribution of energy emitted by a germicidal lamp*

Wavelength (nm)	Percent relative emission within the region[a]		
	248–435 nm	248–365 nm	248–313 nm
248	0.1	0.1	0.1
254	88.5	95.2	97.4
265	0.1	0.1	0.1
280/289	0.1	0.1	0.1
297	0.3	0.3	0.3
302	0.2	0.2	0.2
313	1.7	1.8	1.9
334	0.1	0.1	–
365	1.9	2.0	–
405	2.0	–	–
435	5.0	–	–

[a] The energy in a particular wavelength is expressed as the percentage of either the total emission in the 248–435 nm region, or emission in the UV region only (248–365 nm), or emission in the far UV region (248–313 nm). Figures are calculated from data obtained through the courtesy of General Electric Corp. for the lamp G 30T8.

tensity of 185 nm radiation would greatly depend on the individual lamp and the experimental conditions, it is generally recommended that ozone-free lamps be used. Nevertheless, the strong absorption of 185-nm photons in air makes it relatively safe to irradiate with an ozone-producing lamp at a distance of 25–30 cm, where the 185-nm intensity is several orders of magnitude lower than at the lamp surface.

1.4.5 Solar ultraviolet radiation

Sunlight is the only source of UV radiation to which many organisms are exposed in their natural environment. Hence the study of solar UV effects is of general biological interest and deserves attention. Although the experimental use of solar radiation encounters greater difficulties than the use of a technical UV source in the laboratory, our increasing knowledge in the field of UV photobiology makes a correct interpretation of sunlight effects feasible. A more detailed description of the biological effects of solar UV radiation will be given in Chapter 11.

2 The study of biological UV effects

2.1 Biological objects

Investigations of biological UV effects have largely employed microorganisms, in particular *bacteria* and *bacterial viruses* (often called bacteriophages, or phages), and to lesser extent *yeasts*, other *fungi, animal viruses,* or *plant viruses*. More recently, *mammalian and other vertebrate cells* in culture have become important objects for UV studies, partly because of the awareness of UV carcinogenesis (Chapter 12) and other adverse UV reactions with the human skin (Section 11.4), and partly because of the interest in mammalian repair systems in general, which are best investigated after UV damage (Chapters 7 and 8). Investigations of UV effects in larger organisms, notably protozoa, insects, and spermatophytic plants, are widely scattered in the literature. Because such work encounters greater difficulties regarding irradiation techniques and dosimetry than the microbial work, the results and interpretations have been more open to criticism.

The choice of a biological object for UV studies depends on several things. A decisive factor, besides the scientific background and experience of the investigator, is often whether the purpose of the study is of basic-scientific or of applied-scientific character. In the latter case, use of a particular organism is sometimes dictated by the nature of the problem, irrespective of the suitability of the organism for experimental work. In contrast, UV studies carried out to obtain basic scientific insights will generally employ organisms most appropriate for the experimental approaches envisaged.

In UV photobiology, as in any other area of research, a large amount of knowledge originates from *model systems*. When a suitable organism has been extensively studied through a considerable period of time and thus many of its characteristics are well established, further work is greatly facilitated by the existence of appropriate techniques and the availability of a great variety of mutant strains. Consequently, it is most reasonable to employ the same organism, rather than any other, for answering newly arising scientific questions, unless there are compelling reasons to do otherwise. One reason to use sometimes rather different objects is to get an indication as to whether the scientific knowledge gained with the model species is of general validity. But in this brief introduction into the biological UV effects we will particularly concentrate upon those few objects that have been commonly used and from which most of our present knowledge has been derived.

The preference for bacteria, phages, and some other microorganisms in UV-photobiological studies is well justified. Their very high reproduction rate and small size, the ease and inexpensiveness of obtaining large individual numbers, make them ideal objects for quantitative work. For example, 1 ml liquid culture of the bacterium *Escherichia coli* may contain 10^9 to 10^{10} cells, which within a day could produce the same number of macroscopically visible colonies. Appropriate selection techniques after UV irradiation may permit recognition of a few colonies with altered properties, resulting from a few altered cells among the vast number plated. Another advantage of using bacteria and viruses in UV-photobiological research is their ready accessibility for far-UV radiation. Though they consist of a high proportion of strongly absorbing material, their small diameter (usually below 2 μm) permits the radiation to reach most parts of an individual cell equally well. In contrast, larger biological objects with a diameter of tens or hundreds of μm, though appearing small to the human eye, may be merely affected on the surface, and critical components inside may absorb little, if any, UV.

Of further considerable importance is the fact that some bacterial and viral nucleic acids can be irradiated in vitro, post-treated in various ways, and yet their biological function (as well as their photochemical alterations) tested. Examples are bacterial transforming DNA from *Haemophilus influenzae*, *Bacillus subtilis*, and other species, as well as various infective viral nucleic acids. With transforming DNA one considers essentially a small region of the macromolecule carrying a particular genetic marker that enters the competent (transformable) cell and is integrated into the recipient genome. In work with infective virus DNA one looks usually at the whole nucleic acid molecule, which under appropriate experimental conditions behaves like the complete virion, unless it is affected by the radiation. The experimental opportunities offered by such infectious nucleic acids systems will become evident in later sections.

As has been true for other areas in molecular biology, the use of bacteria and viruses, in particular phages, as model systems has played an essential role in the development of UV photobiology. Knowledge obtained from their study can now be successfully applied to problems encountered with more complex biological systems and structures. Among bacteria, by far the best-investigated species is *Escherichia coli*. In no other bacterium is the chromosome map known in such detail, or are so many different types of mutant derivatives available. Most UV studies with bacteria have been made using *E. coli* strains, and nowhere else is there as much detailed knowledge about repair processes after UV damage. Therefore, omission of specifying a species with which certain results are obtained usually implies that it is *E. coli*.

E. coli serves as a host for many types of phages. Some of them have become model viruses for UV photobiology and several other areas of molecular biology. Phage T4 is certainly the best-investigated type of complex virulent phage, and the same is true for λ as a temperate phage. The information obtained about their UV effects has often been the basis for UV studies in other viral systems, particularly animal and plant viruses, where the experimental procedures involve greater difficulties. Similarly, the analysis of UV phenomena in *E. coli* cells, which has clarified the causal chain from UV absorption via photoproduct formation and repair processes to the finally observed biological effect, has been the basis for understanding the UV photobiology of other prokaryotic as well as eukaryotic cells.

2.2 UV radiation as experimental tool and environmental factor

There are two fundamental aspects to the application of UV radiation in biological experiments: (1) UV can be used as an *experimental tool*, which helps gaining insights into biological functions; (2) UV radiation, in the form of sunlight, constitutes a *natural environmental factor*, whose impact on biological systems (in a positive or negative sense) is of great interest. Each of these two aspects alone would have justified the efforts made to understand biological UV effects. Although the corresponding two emphases in the direction of research differ in principle, they complement each other, and the achievements and interpretations of experimental results often overlap.

As an experimental tool. The principle of using UV radiation as a tool is essentially this: Application of an external factor (UV) causes in cells or other biological systems temporary or permanent alterations that can be identified. They result from photochemical reactions in biomolecules, whose functioning can be assessed relative to the quality and quantity of damage inflicted in them. One might think that many other physical or chemical agents could serve the same purpose, but they usually lack some of the advantages associated with UV radiation. Ultraviolet light acts predominantly through photochemical alterations in DNA. Although other cell components may be affected, DNA is of prime importance because of its unique role as genetic material and its very high UV sensitivity (see Chapter 3). Within DNA, pyrimidine bases are essentially affected, with predominant formation of one group of photoproducts: the 5,6-cyclobutyl dipyrimidines (*pyrimidine dimers*). Quantitative assessment of their formation as a result of irradiation, and of their disappearance as a result of repair, is achieved by conventional chromatographic methods. Comparative studies in vitro with biologically active DNA (see Chapter 5) are valuable for complementing results obtained with whole cells or viruses.

Because of the central role played by DNA within the cell, these primary photo-effects can have quite a variety of biological consequences, in particular lethality, mutation induction, delay of growth or cell division, derepression, or enhanced genetic recombination. Even in heavily UV-damaged

DNA, the capability of phenotypic expression of specific single genes, their relative size, or the polarity in their transcription can be determined. The quantity of the damaging agent is usually better defined in studies with UV radiation than in studies with chemical reagents. In the latter case the time factor is more difficult to control, the substances themselves may undergo alterations in the cell, or their concentration at the critical location may differ considerably from the outside concentration.

As an environmental factor. The only UV radiation in our natural environment to which most organisms, including humans, are exposed at varying degrees, is in the form of sunlight. It contains harmful wavelengths at intensities lower than those applied experimentally, but the duration of solar exposure is usually much longer so that the effects are comparable. Therefore, knowledge acquired in laboratory experiments on UV action and repair is highly valuable for an assessment of the interaction of solar UV with living matter. This subject has been approached both theoretically and experimentally ever since the germicidal effectiveness of sunlight was discovered in the latter part of the nineteenth century. But only recently, after laboratory results provided evidence for specific UV interactions with DNA and for the occurrence of extensive repair, has this kind of work become readily interpretable.

Another type of environmental (though nonnatural) UV exposure of particular concern for humans involves the use of fluorescent lamps for room illumination. Although the glass envelope absorbs all far-UV emission, the commonly used daylight or cool white lamps radiate appreciably in the 313, 334, and 365 nm mercury lines. Much stronger emission at these wavelengths is typical of the so-called blacklight lamps, sometimes used in rooms to cause fluorescent effects. Considerations regarding UV radiation as an environmental factor also prevail in problems of applied-scientific or technical nature, that is, where UV sources are used for the inactivation of microbes in the process of sterilization of apparatus, air, water, or other fluids.

2.3 Quantitative determination of UV radiation energy

UV photobiology has always been a quantitative science. Although the first recognition of an irradiation effect may be based on qualitative criteria, its subsequent investigation inevitably concerns the *quantity* at which it is produced. This is particularly important where UV-induced effects likewise occur spontaneously, though at lesser frequency. Only where the differences between spontaneous and UV-induced effects are very large can they be regarded as if they were of a qualitative nature.

We can reasonably expect that, for a defined wavelength of UV radiation, the quantity of biological effects, corrected for spontaneous occurrences, increases with the amount of radiation energy absorbed. Evaluation of the *dose-effect statistics* has been among the principal approaches toward under-standing UV-photobiological phenomena, and has led to many important con-clusions. Therefore, accurate determination of the radiation quantities ap-plied is vital in any UV-photobiological experiment. It was common in the literature of the 1950s and earlier to express radiation quantities in *relative* terms, for example, in units of "time of UV exposure" (under otherwise fixed conditions). Even with detailed description of the irradiation condi-tions (installation and type of the lamp, distance from the sample, etc.), it was then difficult to compare quantitatively the results obtained in different laboratories. However, within one laboratory the use of relative units enabled the experimentalist to establish the dependence of UV effects on the geno-type of organisms, pre- and postirradiation treatment, or other experimental parameters.

Many valid conclusions had been drawn from such experiments, based solely on relative determination of the UV quantity. Absolute measurements of UV doses published in those times should be considered with caution, as results obtained by different authors often differed to an unexpected extent. An urgency for establishing biological results in terms of absolute UV doses did not prevail until the early 1960s, when fundamental discoveries in the UV-photochemistry of nucleic acids called for quantitative comparisons with photobiological results. Today, most laboratories engaged in this kind of work have adequate equipment for measuring UV quantities, and one can reasonably expect that the stated values are accurate within ±20 percent limits.

2.3.1 Incident energy

The most commonly measured quantity of UV radiation in a biological ex-periment is the energy incident per unit area normal to the beam at the position of the biological material. The term for this quantity is *energy fluence.*[1] Until recently it was expressed in terms of [erg · mm^{-2}] or [erg · cm^{-2}], but in view of the general adoption of m-kg-sec units (instead of cm-g-sec units) it has become common to express the fluence in [J · m^{-2}], a practice even enforced by some journals. Since 1 Joule = 10^7 ergs, and 1 m^2 = 10^6 mm^2 = 10^4 cm^2, the following relation holds:

$$1 \text{ J} \cdot \text{m}^{-2} = 10 \text{ erg} \cdot \text{mm}^{-2} = 10^3 \text{ erg} \cdot \text{cm}^{-2}$$

Thus, fortunately, any conversion of data from one of these units into another requires only a shift in the decimal point.

To calculate the *photon fluence*, which expresses the number of photons incident per unit area, we simply divide the energy fluence by the energy of a single photon at the wavelength applied, which is given by equation (1.2) or (1.3), using compatible units.

To determine the energy fluence it is customary to measure the *fluence rate* (i.e., the fluence per unit time), provided it is kept constant, and to multiply it by the time of exposure. In the common case of irradiation with a germicidal lamp, a *photovoltaic* meter is conveniently used for this purpose, a popular version of which has been described by Jagger (1961). It consists of a photovoltaic cell, covered by a UV-transparent optical filter that absorbs almost all visible light. The current created in the cell by the UV quanta passing the filter, which is read on the scale of a microampere meter, is proportional to the fluence rate. Therefore, by multiplication with a constant factor (determined by calibration of the instrument) one obtains the fluence rate in $[J \cdot m^{-2} \cdot sec^{-1}]$ or $[W \cdot m^{-2}]$.[2] The fluence rate should always be determined exactly at the location of the exposed sample. Therefore, it is advantageous to irradiate at a sufficient distance from the radiation source to minimize errors resulting from slight differences between the location of the sample and that of the measuring device.

Calibration of the photovoltaic meter can be carried out in two ways: (1) by comparison with a *calibrated thermopile*, where the heat created by photon absorption causes a measurable potential difference; and (2) by *actinometry*, for example, measurement of the amount of Fe^{2+} produced by UV absorption in a potassium ferrioxalate solution, for which accurate data concerning absorption and quantum yields are available. Space limitations prevent detailed discussion of these methods; the reader is referred to the book by Jagger (1967). It is advisable to use both methods and to average the results; the degree of consistency between the two measurents is suggestive of their reliability. For measurements of fluence rates at several different wavelengths (e.g., from a monochromator), a thermopile is better suited than a photovoltaic meter, because the latter requires calibration for each wavelength.

Besides these conventional physical or photochemical methods of fluence determination, *biological systems* of known UV sensitivity can serve the same purpose. Virulent bacteriophages, such as phage T2 or T4, are a good choice. The slope of their UV survival curves (see Figure 4.5) is quite reproducible; it is independent of the age, titer, or other particularities of a stock preparation, and only in exceptional cases depends on the type of host cells. Once the quantitative response of such a biological system has been established by a reputable laboratory, it can be applied elsewhere for calibration purposes without the necessity of using physical or photochemical methods. Where it is experimentally feasible, the phage can be directly added to a sample of material to be irradiated. Determination of the phage survival within the sample thus provides a direct measure of the fluence applied, permitting detection of very small differences in UV sensitivities in comparative experiments.

2.3.2 Absorbed energy

Determination of the energy fluence applied to the sample does not tell us how much of the radiation energy is actually absorbed by the biological material or, more specifically, by particular components of it. For a discussion of absorbed energies we will look at several examples of typical experimental arrangements. The degree of complexity encountered in making estimates of the energy absorbed depends greatly on the objects and the manner in which they are exposed to the radiation. Therefore it is important, both for the planning of experiments and for their quantitative evaluation, to understand – in principle – the factors involved. The few examples that we have space for here may serve as an introduction, and, it is hoped, enable the reader to extrapolate to more complex situations.

Light absorption in homogeneous solutions follows relatively simple laws, as will be seen below. However, in UV-photobiological experiments this simple situation is met only occasionally, when free nucleic acids or other molecules of biological importance are irradiated in solution. More often one irradiates suspensions of whole cells or virus particles, which can be considered dense "packages" of absorbing molecules, separated from each other by the suspension liquid. If the liquid itself does not absorb UV radiation (which is usually the case if the liquid is an inorganic buffer solution or a glucose-salts medium), the concentration of the suspended biological units and the thickness of the exposed layer determine whether the sample is essentially *transparent*, *opaque*, or anything between these extremes (*semitransparent*).

Results obtained under any of these experimental conditions are relatively easy to interpret, provided that (1) all particles in the suspension are alike, and (2) their composition permits a photon to pass through them with a high probability (say >0.5) without being absorbed. Greater difficulties in the interpretation of data are encountered with biological units absorbing incident photons with high probability, since in that case considerably more photons react with outer than with inner portions of the units. This happens with large cells, such as ciliate protozoans, or multicellular planktonic organisms. An extreme case is the irradiation of large organisms, where by necessity all UV radiation is absorbed beneath the surface, for example in the *human skin*. Then the likelihood that a cell absorbs UV radiation depends solely on its (fixed) location, decreasing drastically with its distance from the exposed surface. Estimates of radiation absorbed at different depths can be readily obtained, but interpretation of the macroscopic reactions displayed by the whole tissue can be difficult.

Homogeneous samples

Let us first consider the simple case of irradiation of a solution of nucleic acid molecules, for example, infective viral DNA. If it is exposed in a uniform layer to a parallel beam of monochromatic UV radiation at a fluence rate I_0 [W · m^{-2}], the fluence rate of the unabsorbed radiation emerging after passage through the layer is expressed by

$$I = I_0 \cdot 10^{-A} \quad \text{or} \quad I = I_0 e^{-A \ln 10} \tag{2.1}$$

where A is defined as the *absorbance* (or *optical density*) and ln is the natural logarithm, or logarithm to the base e (where $e = 2.71828\ldots$). If the broadside area of the solution, S, is exposed to the fluence F, which is the product of fluence rate and the time, $I_0 t$, we obtain from equation (2.1) for the energy absorbed:

$$E_{abs} = (I_0 - I)\, tS = I_0 t\, (1 - e^{-A \ln 10})\, S = F(1 - e^{-A \ln 10})\, S \qquad (2.2)$$

and for the number of photons absorbed:

$$P_{abs} = E_{abs}/h\nu \qquad (2.2a)$$

The use of the equality $F = I_0 t$ for the derivation of equation (2.2) indicates that in our example the total energy, regardless of the fluence rate and the exposure time, determines the effect of interest. Therefore, this case is said to follow the *Bunsen-Roscoe reciprocity law*.

The nearly transparent case. Calculation of E_{abs} is easiest for solutions that are either nearly transparent or virtually opaque. In the nearly transparent case, where $A \ll 1$, we can approximate the expression $1 - e^{-A \ln 10}$ in equation (2.2) by $A \ln 10$, or $2.3A$, and obtain

$$E_{abs} = 2.3 FAS \qquad (2.3)$$

By the *Lambert-Beers law*, $A = \epsilon c x$, where ϵ is the absorptivity (or extinction coefficient) of the absorbing material at a given wavelength, c is its concentration, and x is the layer thickness. Hence

$$E_{abs} = 2.3 F \epsilon c x S \qquad (2.3a)$$

Since cxS equals the total quantity of absorbing material (or q), the energy absorbed per unit quantity of absorbing material is expressed by

$$D = E_{abs}/q = 2.3 \epsilon F \qquad (2.3b)$$

Thus D, called the *absorbed dose*, is in the near-transparent case independent of the concentration and the geometrical shape of the sample, but varies with ϵ. We will see that nucleic acid solutions as well as suspensions of viruses and small cells are often irradiated under conditions where equation (2.3b) holds.

The units characterizing ϵ, c, and x must be compatible, because the derivations require that their product (A) be dimensionless. If, as usual, ϵ is expressed in terms of *molar* absorptivity (or *molar* extinction coefficient), having dimensions [liter $\cdot$ mole^{-1} $\cdot$ cm^{-1}], c must be expressed in [mole $\cdot$ liter^{-1}] and x in [cm]. In order that $D = 2.3\epsilon F$, q must then be expressed in [mole] and F in [J $\cdot 10^{-3} \cdot$ cm^{-2}]. On the other hand, if the fluence is measured as usual in [J $\cdot$ m^{-2}], the absorbed dose is expressed by

$$D[\text{J} \cdot \text{mole}^{-1}] = 2.3 \times 10^{-1}\, \epsilon[\text{liter} \cdot \text{mole}^{-1} \cdot \text{cm}^{-1}]\, F[\text{J} \cdot \text{m}^{-2}]$$
$$(2.3c)$$

Other compatible combinations of units may be derived from equation (2.3b) by appropriate changes in the numerical constant.

The nearly opaque case. If virtually all UV energy entering the solution is absorbed ($A > 2$), it follows from equation (2.2) that

$$E_{abs} = I_0 tS = FS \qquad (2.4)$$

If the solution consists of a single absorbing component and is well enough stirred to expose all molecules equally, we can write (because $D = E_{abs}/q$, and $S = q/cx$)

$$D = F/cx \qquad (2.4a)$$

Thus, in contrast to the near-transparent case, the energy absorbed per unit quantity of material depends on its concentration and the thickness of the irradiated layer, but is independent of ϵ. This is often convenient for photochemical studies on pure substances. In solutions containing several absorbing components, the fraction of the total energy absorbed by each of them varies with the wavelength as their relative extinction coefficients vary.

Molecules in a stirred opaque solution absorb essentially while they are near the surface, but are temporarily shielded deeper inside the solution. The resulting intermittent irradiation is of no consequence for nonmetabolizing systems, such as a DNA solution, where only the total energy absorbed determines the effect. However, it can be important for a suspension of cells, where the time pattern of the occurrence of radiation damage may become a significant consideration (see Section 8.2).

The intermediate (semitransparent) case. A condition intermediate between near transparency and opacity is represented by a solution that absorbs a measurable fraction of the incident photons but lets the remainder pass. If the solution is well stirred so that all molecules are statistically equally exposed to the radiation, such a case can be treated, for the purpose of our calculations, as a near-transparent solution receiving a lower fluence. Thus the fluence absorbed is expressed by an equation resembling equation (2.3b) or (2.3c), where the factor 2.3 is simply replaced by a smaller, absorbance-dependent factor:

$$D = E_{abs}/q = \left[\frac{1 - e^{-A \ln 10}}{A} \right] \epsilon F \qquad (2.5)$$

or

$$D[J \cdot mole^{-1}] = \left[\frac{1 - e^{-A \ln 10}}{A} \right] \times 10^{-1} \epsilon [liter \cdot mole^{-1} \cdot cm^{-1}] \, F[J \cdot m^{-2}] \qquad (2.5a)$$

(Notice that for $A \ll 1$ the expression in the bracket equals $\ln 10$, or 2.3.)

Table 2.1 *Fluence correction factors for partially transmitting samples*

Sample transmission T	Sample absorbance $A = -\log T$	Morowitz correction factor $M(A) = \dfrac{1 - e^{-2.3A}}{2.3A}$
1.00	0	1
0.90	0.046	0.95
0.80	0.097	0.90
0.70	0.155	0.84
0.60	0.222	0.78
0.50	0.301	0.72
0.40	0.398	0.66
0.30	0.523	0.58
0.20	0.699	0.50
0.10	1.000	0.39
0.05	1.301	0.32
0.02	1.699	0.25
0.01	2.000	0.215

Accordingly, as pointed out by Morowitz in 1950, the average fluence actually received by the molecules in the solution equals the fluence at the surface, multiplied by the factor $M(A) = (1 - e^{-A \ln 10})/2.3A$, which is called the *Morowitz correction* factor. It is evident that one arrives at equations (2.5) and (2.5a) by multiplying, respectively, equations (2.3b) and (2.3c) with $M(A)$. It is convenient to look up $M(A)$ on a graph or in a table (see Table 2.1) after measuring the transmission of the sample.

Particulate samples

For simplicity, the previous equations were derived for homogeneous samples, that is, molecules in solution. However, with the exception of studying UV-effects on biologically important molecules in vitro, the situation encountered in actual UV-photobiological experiments is generally more complex. If suspensions of viruses, single cells, or cell organelles are to be irradiated, the following considerations apply: (1) The suspension is a random assortment of absorbing packages, rather than a uniform solution of molecules. (2) The absorbing packages are optically inhomogeneous, and their average index of refraction usually differs from that of the surrounding fluid. Both of these conditions cause significant light scattering; its extent depends on small details of the particle structure because the particles have dimensions of a similar order of magnitude to UV wavelengths. (3) The suspended units often contain other absorbing molecules besides those responsible for the biological effect of interest.

All three of these considerations complicate computation and experimental determination of the energy absorption required for the biological effect. However, much of the simplicity found with homogeneous solutions still holds for experimental conditions where (a) the entire particle suspension is sufficiently dilute and exposed in a thin layer as to be essentially transparent, (b) every part of each individual particle is itself nearly transparent, and (c) the total scattering is not large. If this is the case, individual components absorb radiation independently of one another and irrespective of their own precise location within the particle. In effect, the suspension is acting like an array of little homogeneous, near-transparent solutions, within each one of which equation (2.3b) governs. The validity of this equation for homogeneous samples required that $A \ll 1$. We apply the same requirement to each particle and to each portion of it. Because $A = \epsilon c x$ and $cx = q/S$ (where S is now the broadside area of an individual particle, or portion of a particle, and q the corresponding amount of absorbing material), the requirement becomes $\epsilon q/S \ll 1$. As always, compatible units, giving ϵq the same dimensions as S, must be employed in the calculations.

The degree of transparency of individual particles in solution can be determined from the measured absorption of a suspension of randomly distributed particles at known concentration, provided that both light scattering and the so-called *sieve effect* are taken into consideration. The latter is a consequence of the statistical particle distribution, which permits a fraction of the incident photons to pass the fluid without ever encountering a particle. Thus the absorbancy ρ_c of an individual can be approximated by

$$\rho_c = \log \frac{0.434 P}{0.434 P - \rho_s} \qquad (2.6)$$

where ρ_s is the absorbance of the whole suspension of particles, already corrected for light scattering, and P is the number of layers (or the fraction of a layer) of such particles that would be obtained if all particles were lined up without space between them on the broadside area of the irradiated volume.

A rough estimate of the absorbance of individual particles can also be obtained by calculation if, besides the particle size, the amount and absorption properties of the major component substances are known (see example in Section 2.3.3). If particles are fairly transparent, but the suspension is so concentrated or the layer so thick that total absorption becomes appreciable, the Morowitz correction factor can be applied as in the case of homogenous solutions. Such situations are not seldom encountered when the experiment requires the irradiation of very many individual biological units. If they move around freely, the suspension should be stirred in order to expose all individuals to the same *average* quantity of radiation.

Rather than using the Morowitz correction, one may as well experimentally determine the correction factor for the conditions specifically applied. This requires measuring, besides the fluence-effect curve for such a semitransparent particulate sample, the analogous curve for a near-transparent sample of the same kind (obtained by lowering the particle concentration). The two curves, resulting from exposure of the samples to the same fluence rate and for the same times under identical experimental conditions, should be of similar shape. They are expected to become congruent if each irradiation time for the semitransparent sample is multiplied by a constant factor <1 (namely, the correction factor), which can be graphically determined.

The concentration of particles (cells, viruses, molecules, etc.) in a suspension is usually described in terms of their number, n, per unit volume. In this case the absorbance of particles is conveniently expressed by their cross section for absorption, σ. If equation (2.1) is written in the form $I/I_0 = e^{-2.3\epsilon cx}$ (where $\epsilon cx = A$ and $2.3 \approx \ln 10$), the absorption-characteristic term $2.3\,\epsilon c$ of a solution can be replaced by that for a particulate suspension $n\sigma$. Thus:

$$I/I_0 = e^{-\sigma nx} \tag{2.7}$$

The coefficient σ has the dimensions of an area, thereby making the exponent dimensionless. Its meaning is easily visualized by picturing the incident photons as randomly fired bullets, where the average number absorbed per particle equals the number that would accidentally strike an area equaling the cross section σ. If the fluence is expressed in terms of incident photons per unit area ($F_p = F/h\nu$), the number of photons absorbed per particle in a sufficiently dilute suspension is given by

$$P_{abs} = \sigma F_p \tag{2.8}$$

Determination of energy absorption becomes more complicated when *polychromatic UV radiation* is used, where both the fluence and the absorbance differ for different wavelength bands. Although from a theoretical point of view such irradiation is less desirable than irradiation at a single defined wavelength, practical reasons (as for example the study of effects of natural sunlight) might be of overriding concern.

For samples that are nearly transparent at all wavelengths of interest, equation (2.3b) can be written as an integral covering the wavelength interval from λ_{min} to λ_{max}:

$$D = 2.3 \int_{\lambda_{min}}^{\lambda_{max}} \epsilon(\lambda) F(\lambda)\, d\lambda \tag{2.9}$$

Consequently, the number of photons absorbed is expressed by:

$$P_{abs} = \frac{D}{h\nu} = \frac{2.3}{h\nu} \int_{\lambda_{min}}^{\lambda_{max}} \lambda \epsilon(\lambda) F(\lambda)\, d\lambda \tag{2.9a}$$

If curves for $\epsilon(\lambda)$ and $F(\lambda) = I_0(\lambda)\, t$ versus λ are available, such integrals may be evaluated graphically. Alternatively, numerical values for ϵ and I_0 tabulated for discrete wavelength intervals can be used for the calculation.

2.3.3 Numerical examples

A few numerical examples will now show how the considerations in Sections 2.3.1 and 2.3.2 and the equations derived from them apply to actual experimental situations.

Homogeneous samples

Example 1. A thin layer of an aqueous solution of infectious HP1 phage DNA molecules (MW = 20×10^6 daltons) is exposed to $10\ \mathrm{J} \cdot \mathrm{m}^{-2}$ ($= 10^{-3}\ \mathrm{J} \cdot \mathrm{cm}^{-2}$) UV radiation of 254 nm wavelength. The DNA concentration is sufficiently low that the solution is essentially transparent. We wish to calculate how many photons are absorbed per DNA molecule.

We know that most double-stranded DNAs in the native state show an optical density $A = \epsilon cx$ close to 0.22 at 254 nm for a 10 μg/ml solution 1 cm thick, which is equivalent to $\epsilon = 2.2 \times 10^{-2}$ [ml $\cdot \mu\mathrm{g}^{-1} \cdot \mathrm{cm}^{-1}$]. Thus, according to equation (2.3b) we obtain:

$$D = 2.3 \times 2.2 \times 10^{-2}\, [\mathrm{ml} \cdot \mu\mathrm{g}^{-1} \cdot \mathrm{cm}^{-1}] \times 10^{-3}\, [\mathrm{J} \cdot \mathrm{cm}^{-2}]$$
$$= 5.06 \times 10^{-5}\ \mathrm{J} \cdot \mu\mathrm{g}^{-1}$$

Because 1 dalton = $1.66 \times 10^{-18}\ \mu$g, a DNA molecule of 20×10^6 daltons has a weight of $3.32 \times 10^{-11}\ \mu$g, and the energy absorbed by such a molecule is thus $5.06 \times 10^{-5} \times 3.32 \times 10^{-11}$ or $1.68 \times 10^{-15}\ \mathrm{J} \cdot \mathrm{molecule}^{-1}$. According to equation (1.2) the energy at 254 nm is $7.83 \times 10^{-19}\ \mathrm{J}$ per photon, so that the absorption per DNA molecule is, on the average, $(1.68 \times 10^{-15}\ \mathrm{J} \cdot \mathrm{molecule}^{-1})/(7.83 \times 10^{-19}\ \mathrm{J} \cdot \mathrm{photon}^{-1}) = 2.15 \times 10^3$ photons $\cdot$ molecule^{-1}.

Example 2. For the same experimental situation as in example 1 we wish to calculate the number of photons absorbed per DNA *nucleotide*, which could be of interest in comparison with photochemical events in model systems. This number can be easily obtained from the results of example 1 by dividing 2.15×10^3 by the number of nucleotides per DNA molecule (i.e., 6.7×10^4 in the case of HP1 DNA). The result is 3.2×10^{-2} photon absorbed per nucleotide.

If only ϵ of the solution is known but not the size and the number of molecules, the following calculation will give the same answer. According to equation (2.3b):

$$D\,[\mathrm{J} \cdot (\mathrm{nucleotides})^{-1}] = 2.3\,\epsilon\,[\mathrm{cm}^3 \cdot (\mathrm{nucleotides})^{-1} \cdot \mathrm{cm}^{-1}] \cdot F\,[\mathrm{J} \cdot \mathrm{cm}^{-2}]$$

The extinction coefficient of DNA (or other nucleic acids) is often expressed by ϵ_p, the molar extinction coefficient when its concentration is given in moles of DNA phosphorus per liter, which is about 6700 liter · mole^{-1} · cm^{-1}. Because there is one phosphorus atom per nucleotide, ϵ_p is at the same time the molar extinction coefficient of the DNA nucleotides in situ in the nucleic acid. The term 6700 liter · mole^{-1} · cm^{-1} equals (6700 · $10^3/6.02 \times 10^{23}$) cm^3 · (nucleotide)$^{-1}$ cm^{-1}, so that

$$D = \frac{2.3 \times 6.7 \times 10^6 \times 10^{-3}}{6.02 \times 10^{23}} \quad \text{or} \quad 2.56 \times 10^{-20} \text{ J/nucleotide}$$

For a photon energy at 254 nm of 7.83×10^{-19} J, the average number of photons absorbed per nucleotide is therefore about 3.2×10^{-2} or about 1/30. Thus, although over 2000 primary excitations have occured in this large molecule, the majority of the bases have absorbed no photons at all.

Example 3. We now consider a DNA at high concentration (100 μg/ml), stirred in a layer 1 cm thick, for which the absorbance $A = 2.2$. We wish to know the fluence required to produce the same average energy absorption *per molecule* as was obtained by 10 J · m^{-2} on the transparent DNA solution.

We had calculated in example 1 that at a fluence of 10 J · m^{-2} the transparent solution absorbs 5.06×10^{-5} J · μg^{-1}. This means, by equation (2.4a), that

$$D = \frac{F}{cx} = 5.06 \times 10^{-5} \text{ J} \cdot \mu\text{g}^{-1}$$

Thus the required fluence F equals 5.06×10^{-5} [J · μg^{-1}] · 100 [μg · cm^{-3}] · 1 [cm] = 5.06×10^{-3} J · cm^{-2}, or 50.6 J · m^{-2}, which is more than five times the 10 J · m^{-2} required to cause the same absorption per molecule in the near-transparent case. On the other hand, the total number of photons absorbed per unit incident fluence is in the same cuvette volume much greater at the higher DNA concentration than at the low concentration. For a well-stirred semitransparent sample of known absorbance A, the calculation would be made according to equation (2.5).

Particulate samples

Example 4. We wish to calculate the transmission for 254 nm radiation by a T2 bacteriophage, with head dimensions corresponding to a broadside area S of roughly 5×10^{-11} cm^2, and containing about 2.2×10^{-10} μg DNA and the same amount of protein. Nucleic acid absorbance at 254 nm is much stronger than protein absorbance so that the latter can be ignored in our calculation. Because for native double-stranded DNA $\epsilon = 2.2 \times 10^{-2}$ ml ·

$\mu g^{-1} \cdot cm^{-1}$, the absorbance of the whole phage $A = \epsilon q/S$ is approximately 0.1, and the transmission $I/I_0 = 10^{-A}$ is about 80 percent. Clearly for any smaller structure in the particle, having composition not too different from the average of the whole, A would be less. Thus the condition $A \ll 1$ is met, and irradiation of such particles in dilute suspension should result in the same energy absorption per unit mass of DNA as if the DNA were in a transparent homogeneous solution, provided the extinction coefficient is similar under these two conditions.[3]

Example 5. A bacterial cell is both larger in size and more complex in composition than a virus particle, and both of these parameters vary with growth conditions. A rough estimate of the transmission of an *E. coli* cell at 254 nm can be based upon the fact that absorption is dominated by nucleic acids (of which at least 80 percent is RNA), which make up about 5 percent (or 5×10^{-14} g) of the total mass of 10^{-12} g of a growing *E. coli* cell. The cell size is about 2×10^{-4} cm in its longest dimension and about 0.8×10^{-4} cm in its shortest. Because the extinction coefficient for cellular RNA is not greatly different from the value we have previously used for DNA, the absorbance of the cell $A = \epsilon q/S$ amounts to $(2.2 \times 10^{-2}$ ml $\cdot \mu g^{-1} \cdot cm^{-1}) \times (5 \times 10^{-8}$ $\mu g)/(1.6 \times 10^{-8}$ $cm^2)$, or 7×10^{-2}, for broadside illumination. For end-on illumination (giving a circular cross section) the figure would be about 0.22. These values correspond to particle transmissions of about 85 and 60 percent, respectively. Because the average S value for all possible orientations of the cell corresponds closer to broadside than to end-on illumination, 85 percent should be closer to the actual figure. Stationary phase cells, which contain less nucleic acids, would be more transparent.

Thus, according to our rough calculation, the transmission of a bacterial cell is still fairly high, but for larger-sized cells (e.g., yeasts, protozoa, mammalian cells), the use of the transparent approximation at this wavelength becomes questionable. For wavelengths below 230 nm, where protein absorbance becomes considerable, closer attention must be given to estimating the penetration of radiation.

3 Interaction of UV radiation with molecules of biological importance

3.1 UV absorption by nucleic acids, proteins, and other biologically relevant molecules

Nucleic acids. The biological effectiveness of UV radiation is primarily due to its absorption by nucleic acids, and to a much lesser extent by proteins and other biologically important molecules. The absorbing components within nucleic acids are the *nucleotide bases*, in DNA usually the purine derivatives adenine (A) and guanine (G), and the pyrimidine derivatives thymine (T) and cytosine (C). RNA normally contains uracil (U) instead of thymine. Figure 3.1 shows that the absorption spectra of the component bases differ somewhat; and since nucleic acids from a variety of organisms may show considerable differences in their base compositions, not all absorption spectra of nucleic acids are alike. Nevertheless, their common features are an absorption maximum in the 260–265 nm region and a rapid decline toward longer wavelengths, which makes absorption measurements above 320 nm rather difficult. As seen in Fig. 3.2, the spectrum has a shallow minimum near 230 nm, before it increases again toward shorter wavelengths.

The absorption characteristics of RNA resemble those of DNA. Nevertheless, UV-induced alterations in RNA are usually of much lesser biological consequence than similar alterations in DNA, because of the uniqueness of the latter molecule as the genetic material. RNA, in the form of either messenger RNA, transfer RNA, or ribosomal RNA, is present in many copies and can be replaced as long as the information for their production, carried by the DNA, is unaffected. Exceptions are those biological systems in which RNA is the only genetic material, as in some animal and plant viruses as well as in a few bacteriophages. Even near-UV radiation, whose absorption by DNA is small or unmeasurable, causes most lethal effects by way of DNA alterations, either directly or through sensitization by other molecules.

Proteins. Absorption of most proteins in the far-UV region from 240 to 300 nm is much lower than that of nucleic acid solutions of equal concentration (see Figure 3.2). The reason is that, in contrast to all the nucleotide residues in DNA, only a few of the amino acid residues of pro-

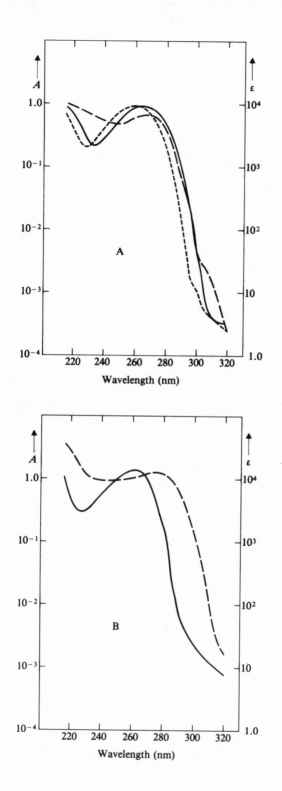

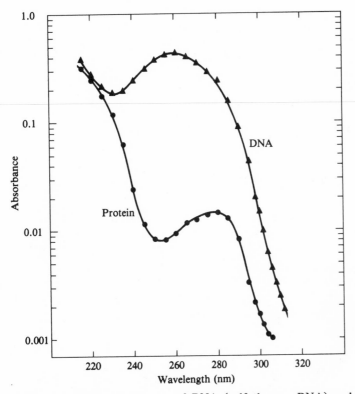

Figure 3.2. Absorption spectra of DNA (calf thymus DNA) and a protein (bovine serum albumin), both at concentrations of 19.3 μg/ml.

teins absorb measurably in this region. Figure 3.3 indicates that in particular the absorption maximum of proteins around 280 nm and the increase below 250 nm are accounted for by tryptophan and tyrosine, although they are uncommon amino acids. In comparison, cystine, cysteine, phenylalanine, and histidine absorb much less, but their contribution may become relevant in proteins containing little or no tryptophan or tyrosine. The high protein absorption below 230 nm, which resembles that of DNA of equal concentration, is primarily due to the absorption properties of the peptide bond. Al-

Figure 3.1. *Panel A:* Absorption spectra of the pyrimidine bases thymine (solid curve), cytosine (dashed curve), and uracil (short-dashed curve) in 10^{-4} molar aqueous solutions. *Panel B:* Absorption spectra of the purine bases adenine in aqueous solution (solid curve) and guanine in 30% NH_4OH (dashed curve) in 10^{-4} molar concentrations. Absorbance for 1-cm light path is shown on the left ordinate scale, the molar extinction coefficient ϵ (in liter $\cdot$ mole^{-1} $\cdot$ cm^{-1}) on the right.

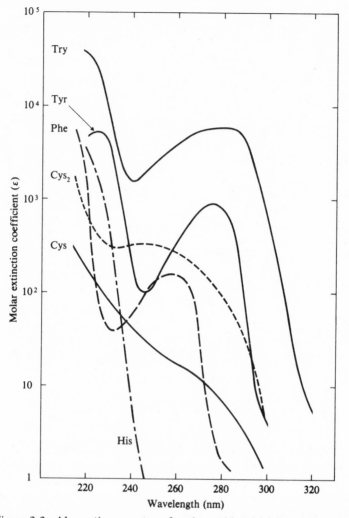

Figure 3.3. Absorption spectra of amino acids with absorption bands in the far UV region. The molar extinction coefficient (ϵ) is plotted as a function of wavelength for tryptophan (Try), tyrosine (Tyr), phenylalanine (Phe), cystine (Cys$_2$), cysteine (Cys), and histidine (His).

though proteins fulfill many vital functions in cells, their UV absorption, compared with that of DNA, is of minor consequence. Most proteins are present in cells in a considerable number of identical copies; therefore, photochemical alterations in only a fraction of them may still permit carrying out the biological function.

Other molecules of biological importance. Biological substances with unsaturated bonds, other than nucleic acids and proteins, may occasionally be involved in the effects of UV radiation. Figure 3.4 shows schematically the positions of UV absorption bands for a number of biologically relevant chromophores. In the near-UV region energy absorption by flavins, porphyrins, steroids, quinones, or carotenoids may have biological consequences, as these substances serve as coenzymes, hormones, or electron

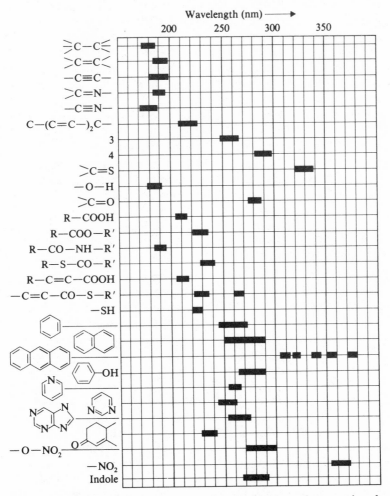

Figure 3.4. Position of UV absorption bands of commonly occurring chromophores. (From J. F. Scott, in: *Physical Techniques in Biological Research*, G. Oster and A. W. Pollister, eds., Vol. 1, Academic Press, New York, 1955, pp. 131–203.)

transport molecules in the respiratory chain. The net results will depend greatly on the biological system and on the particular effect studied.

3.2 Determination of the critical absorbing component: The action spectrum

An action spectrum displays the relative efficiencies of incident photons, as a function of their energies (or wavelengths), for producing a given biological effect. If the *quantum yield*[1] for formation of the relevant photoproducts is approximately constant within the spectral region considered, the action spectrum is expected to resemble the absorption spectrum of molecules responsible for the formation of such photoproducts. The reason for this is that the resulting biological effect should depend only on their amounts, not on the wavelength of the radiation that formed them. Comparison of an action spectrum with absorption spectra of appropriate bio-organic substances might, therefore, indicate which one of them is primarily responsible for the effect.

Long before the function of DNA as the genetic material of organisms was recognized, action spectra indicated that nucleic acids are the predominant UV-absorbing material involved in the inactivation of bacteria and phage. As early as 1928, Gates, working with *Staphylococcus aureus* cells, discovered that the relative bactericidal actions of various UV wavelengths matched their absorbance by the nucleotide bases cytosine, thymine, and uracil. This was confirmed by later data obtained with *Staphylococcus* phages and *E. coli* cells (Figure 3.5). The action spectrum displays maximal effectiveness of wavelengths around 265 nm, with a rapid decrease toward 300 nm and a minimum around 230 nm. Such results contrasted with the earlier belief that photon absorption by the aromatic amino acid residues of proteins is responsible for the destructive effects of UV radiation, and gave the first strong indication for the important role of nucleic acid in UV photobiology.

Within the next decade, various researchers obtained similar action spectra for UV-induced mutagenesis in pollen grains of higher plants and spermatozoids of the liverwort, *Sphaerocarpus donellii* (see Table 9.1). However, the implication that nucleic acid is the genetic material (or an important component of it) was not taken seriously. In the absence of detailed knowledge about the molecular structure of DNA, the tremendous hereditary variability of organisms seemed to be accountable only by the huge variation potential of proteins. The theoretical reservations concerning DNA being the possible genetic material were not overcome until years later, when O. Avery and associates, in 1944, unambiguously demonstrated specific hereditary changes (*bacterial transformation*) by means of DNA.

Establishment of action spectra. The number of molecules M undergoing a certain photochemical reaction in a sample equals the number of

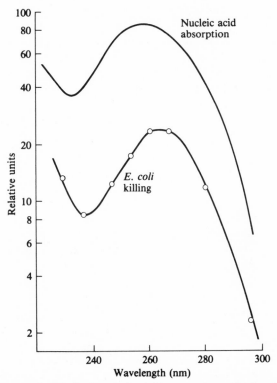

Figure 3.5. Similarity of the action spectrum for inactivation of *E. coli* cells, determined by F. L. Gates, to the absorption spectrum of nucleic acids. (Redrawn from C. S. Rupert, in: *Comparative Effects of Radiation*, M. Burton, J. S. Kirby-Smith, and J. L. Magee, eds., Wiley, New York, 1960, pp. 49–61.)

photons absorbed P_{abs}, multiplied by the quantum yield Φ for the reaction. Because, according to equation (2.2a), $P_{abs} = E_{abs}/h\upsilon = E_{abs}\lambda/h\upsilon$, we can write

$$M = \Phi E_{abs}\lambda/h\upsilon \qquad (3.1)$$

If a virtually transparent sample, for which according to equation (2.3b) $E_{abs} = 2.3q\epsilon F$, is irradiated at two different wavelengths λ_1 and λ_2, with the respective fluences F_1 and F_2 *chosen to produce equal effects* (for example, 90 percent inactivation of a phage population), the following equation holds:

$$(\Phi_1\lambda_1)(2.3q\epsilon_1F_1)/h\upsilon = (\Phi_2\lambda_2)(2.3q\epsilon_2F_2)/h\upsilon$$

which can be rearranged to the proportionality

$$\frac{\epsilon_2\Phi_2}{\epsilon_1\Phi_1} = \frac{1/(\lambda_2F_2)}{1/(\lambda_1F_1)} \qquad (3.2)$$

We see that, for a given effect, the quantity $1/\lambda F$ is at any wavelength proportional to the extinction coefficient ϵ of the component responsible for the effect, multiplied by the quantum yield Φ. Thus at a given wavelength, $1/F$ expresses the relative effectiveness of the incident energy fluence, and – because of the inverse proportionality of photon energy with wavelength – $1/\lambda F$ expresses the relative effectiveness of incident photons. The dependence of the latter expression upon the wavelength λ constitutes the *action spectrum*. In actual experiments it is customary to choose one wavelength (say λ_1) as a reference wavelength, and to determine experimentally the relative values of $1/\lambda F$ for a series of other wavelengths ($\lambda_2, \lambda_3 \ldots \lambda_n$), using equation (3.2).

A plot of $1/\lambda F$ versus λ should resemble the spectrum of the UV-absorbing component responsible for the effect, if the following experimental conditions hold: (a) At each wavelength the effect depends only on the fluence $F = I_0 t$ for different fluence rates I_0 and times of irradiation t (Bunsen-Roscoe reciprocity law); (b) the fluence-response curves at different wavelengths have the same shapes (differing only by a constant factor on the fluence scale), suggesting that the underlying photochemical reactions are the same at the different wavelengths; (c) the suspension and the biological units are virtually transparent at all wavelengths (or suitable corrections can be applied); and (d) the quantum yield Φ is independent of wavelength within the range examined.

Requirements (a), (b), and (c) are sometimes not met in biological systems, and (d) is always an assumption, unless the photochemical reaction is already well characterized – in which case there would be little reason for establishing an action spectrum. The conclusions drawn from action spectra are thus open to criticism. We must be further aware of the fact that excitation energy can be transferred over short distances; therefore, the energy-absorbing molecules are not necessarily those whose photochemical alterations cause the observed biological effect. But if the action spectrum for a certain biological effect resembles the absorption spectrum of some relevant component of the organism, there is at least considerable heuristic value in the suggestion that energy absorption by this particular component is the primary step responsible for the effect. The action spectrum thus constitutes the basis for further experimentation and different approaches, which might settle the matter unambiguously.

Interesting photobiological details may be revealed by action spectra even though the absorbing substance and/or the critical photoproducts for a certain effect are already known. Figure 3.6 compares the action spectrum for inactivation of phage T4 with the absorbance of this phage, corrected for light scattering. Normalization of the two curves at 265 nm, the peak of DNA absorption, indicates that at wavelengths between 270 and 310 nm and below 240 nm the inactivation is lower than expected from the phage

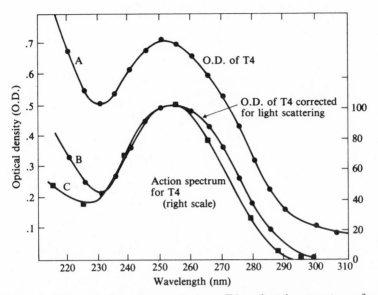

Figure 3.6. Absorption spectrum of phage T4 and action spectrum for its inactivation. *Curves A and B:* T4 absorbance uncorrected, and corrected for light scattering, respectively. *Curve C:* Action spectrum normalized so that its peak at 265 nm coincides with the peak of curve B (% action read on right scale). (From U. Winkler, H. E. Johns, and E. Kellenberger, *Virology 18*, 343, 1962.)

absorbance. This suggests that proteins increase measurably the absorption of the phage in these spectral regions without a commensurate contribution on its inactivation. Figure 3.7 compares the action spectrum for inactivation of phage ΦX 174 with the absorption spectra of the phage particles and the single-stranded phage DNA. At wavelengths below 240 nm the action spectrum differs considerably from DNA absorption, suggesting an increase in the quantum yield for the formation of lethal photoproducts in DNA, perhaps as a result of photochemical DNA-protein interaction.

3.3 Major photoproducts of biological significance

Although both purine and pyrimidine residues in DNA are strong absorbers in the far-UV region (see Figure 3.1), their contributions to biological effects are grossly different. Back in 1952, when very little was known about the UV photochemistry of DNA, Errera pointed out that under identical irradiation conditions pyrimidines undergo photodecomposition at a much higher rate than purines. Since then, all of the major DNA photoproducts identified, including those whose biological relevance has been definitively demonstrated, turned out indeed to be pyrimidine derivatives. Figure 3.8 shows schemati-

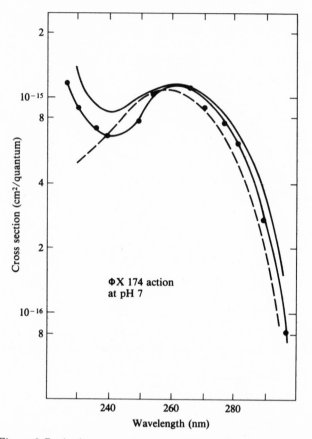

Figure 3.7. Action spectrum for UV inactivation of phage ΦX 174 (filled circles), in comparison with the absorption spectrum of this phage (upper solid line) and the absorption spectrum of the phage DNA (dashed line). (From R. Setlow and R. Boyce, *Biophys. J. 1*, 29, 1960.)

cally the main types of photoproducts in irradiated DNA, which will be briefly discussed in the following sections. They are (a) cyclobutyl type dimers; (b) pyrimidine adducts; (c) the "spore photoproduct," formed only under special conditions; (d) pyrimidine hydrates; and (e) DNA-protein cross-links. For a detailed review of these matters, see Patrick and Rahn (1976).

The absence of purine UV-photoproducts from this list is certainly not fortuitous. Photo-decomposition of naturally occurring purines by 254-nm radiation, as judged by alteration of their absorption characteristics, occurs with quantum yields of the order of 10^{-4}, which is one to two orders of magnitude lower than observed with pyrimidines. Although photoproducts in the purine moieties of DNA could actually be formed more frequently than suggested by these figures (because some of them might have unaltered

Class of photoproduct	Representative photoproduct

Cyclobutane-type dimers	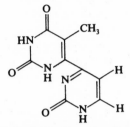 cys-syn thymine-thymine dimer
Pyrimidine adducts	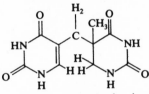 6-4'-[pyrimidin-2'-one] pyrimidine
Spore photoproduct	5-thyminyl-5,6-dihydrothymine
Pyrimidine hydrates	6-hydroxy-5,6-dihydrocytosine
DNA-protein crosslinks	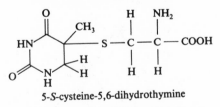 5-S-cysteine-5,6-dihydrothymine

Figure 3.8. Examples of DNA photoproducts formed in UV-irradiated cells.

absorption properties), present knowledge gives no indication for their involvement in the common biological UV effects. For this reason, very little work has been done on purine photoproducts in DNA, even though their possible contribution under unusual conditions cannot be ruled out.

3.3.1 Cyclobutyl dipyrimidines

By far the most important class of UV photoproducts formed in native DNA are the 5,6-cyclobutyl dipyrimidines, commonly called *pyrimidine dimers* (Pyr◇Pyr; see Figure 3.8). In a double-stranded DNA molecule they are formed from two adjacent pyrimidine residues within one DNA strand (*intra*strand dimers), resulting in three types: thymine-thymine (T◇T), thymine-cytosine (T◇C), and cytosine-cytosine (C◇C) dimers. Considering the polarity of the DNA strand, the cytosine-thymine (C◇T) dimer may be distinguished from the T◇C dimer as a fourth type.

If the bases were not restricted in their orientation, each type of dimer could exist in four stereoisomeric forms: (1) cis-syn, (2) cis-anti, (3) trans-syn, and (4) trans-anti, as schematically illustrated in Figure 3.9. However, in native double-stranded DNA only the cis-syn dimer is formed, while in single-stranded DNA the trans-syn type can also occur. It is important to notice that by formation of the cyclobutane ring the 5,6-double bond of the monomeric pyrimidines is lost. Because this double bond accounts for the strong far-UV absorption of these bases, absorption of dimerized pyrimidines is greatly reduced at wavelengths >235 nm (Figure 3.10). But if photon absorption occurs, it causes splitting of the cyclobutane ring with

Cyclobutyl dipyrimidine isomers

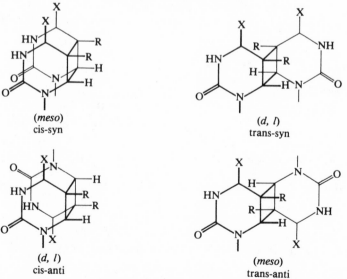

Figure 3.9. The four stereoisomeric forms of cyclobutyl dipyrimidines. (From M. H. Patrick and R. O. Rahn, in: *Photochemistry and Photobiology of Nucleic Acids*, S. Y. Wang, ed., Vol. II, Academic Press, New York, 1976, pp. 35–95.)

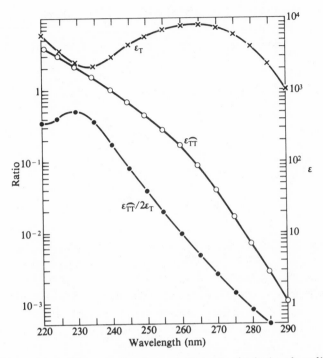

Figure 3.10. Molar extinction coefficients (right-hand ordinate scale) of thymine (ϵ_T) and of thymine-thymine dimers ($\epsilon_{\widehat{TT}}$) in the far-UV region. The ratio [dimer absorbance] : [2 X monomer absorbance] at the different wavelengths is read on the left-hand ordinate scale. (From H. E. Johns, S. A. Rapaport, and M. Delbrück, *J. Mol. Biol. 4*, 104, 1962.)

a high quantum yield, that is, monomerization of the dimer into the component pyrimidines.

The reversibility of dimer formation leads at high UV fluence to a steady state, in which the ratio of dimers/monomers depends on the wavelength of irradiation, the base composition of the DNA, and the particular type of dimer investigated. These considerations are often important for photochemical experiments. Biological UV experiments, with a few exceptions, are carried out at fluences well below those where dimer reversal becomes appreciable. For example, at 254 nm T ◇ T formation in *E. coli* DNA occurs essentially in proportion to the fluence up to at least 10^3 J · m^{-2}. C ◇ T dimerization already slows at considerably lower fluence, but because its production rate is lower than that of T ◇ T, total dimer formation still approximates a linear function for fluences up to 500 J · m^{-2}. This is several times more than applied in most biological experiments. We should be aware, however, that even in the range of fluence proportionality the relative frequencies of the three types of dimers at a given fluence can differ consider-

Table 3.1 *Relative frequencies of pyrimidine dimers in DNA from bacteria with different base-pair ratios*

DNA source	(A + T)/(G + C) ratio	Percentage of dimers formed as		
		T $\diamond$ T	C $\diamond$ T	C $\diamond$ C
Haemophilus influenzae	1.63	71	24	5
Escherichia coli	1.00	59	34	7
Micrococcus luteus	0.43	19	55	26

Note: Obtained by 265-nm irradiation at a fluence of 2×10^2 J · m^{-2}.
Source: R. B. Setlow and W. L. Carrier, *J. Mol. Biol. 17*, 237–54, 1966.

ably when DNAs with different base compositions are compared. This is illustrated in Table 3.1 for DNA from three commonly used bacterial species.

Monomerization of pyrimidine dimers, which Figure 3.10 shows is most favored at wavelengths in the 230–240 nm region, has been an important criterion for proving the biological effectiveness of pyrimidine dimers. Specific experimental data will be discussed in Section 6.2.1.

3.3.2 Pyrimidine adducts

Like the cyclobutane-type dimers, pyrimidine adducts are dipyrimidines formed from adjacent pyrimidine bases within one DNA strand. The best-investigated type is the thymine-cytosine adduct: 5-hydroxy-6-4′[5′-methyl-pyrimidin-2′-one]-dihydrothymine. After acid hydrolysis of irradiated DNA, it is recognized as the dehydrated analogue 6-4′[pyrimidin-2′-one] thymine (see Figure 3.8).

The frequency of photoadduct formation, like that of pyrimidine dimers, depends upon the base composition of irradiated DNA. In *E. coli* DNA, irradiated at room temperature, pyrimidine adducts occur at roughly one tenth the rate of pyrimidine dimers, but the difference is far less at low irradiation temperature (−76°C). Unlike dimers, pyrimidine adducts can not be split photoenzymatically; but their absorption peak at 315 nm permits photochemical destruction at this wavelength at a rate comparable to that of photoenzymatic monomerization of pyrimidine dimers. The latter are not photochemically altered by 315 nm in the absence of photoreactivating enzyme.

The significance of pyrimidine adducts concerning biological effects remains to be clarified. Their possible involvement in UV inactivation of *Streptomyces griseus* is suggested by the photoreactivation action spectrum of this organism, which has a minor peak at 315 nm besides the main peak at 435 nm (see Figure 7.6). While the latter is absent in a mutant lacking photoreactivating enzyme, the minor peak at 315 nm is retained, perhaps reflecting photochemical destruction of pyrimidine adducts in this wavelength region.

3.3.3 Spore photoproduct

A dipyrimidine UV photoproduct in DNA that has definitely been shown to cause lethal effects is the spore photoproduct. It is not found in vegetative bacterial cells irradiated at room temperature, but is the predominant photoproduct in UV-irradiated bacterial spores.[2] This photoproduct is also observed when DNA is irradiated at temperatures between $-50°$ and $-180°C$ (with a maximal yield at about $-100°C$), or in >50 percent ethanol. The important condition for the formation of spore photoproduct is the absence of water, rather than a specific conformation of DNA.

The spore photoproduct has been identified as 5-thyminyl-5,6-dihydrothymine. Its chemical structure (Figure 3.8) shows that one of the two thymine residues still has the 5,6-double bond in the heterocyclic ring and consequently absorbs strongly in the 265 nm region. This photoproduct is obtained at high yield throughout the far-UV range, with a quantum yield exceeding that for cyclobutane-type dimer formation. Unlike cyclobutane-type thymine dimers, the spore photoproduct is not reverted to the two original thymines by either photoenzymatic repair or direct photochemical reaction. But *Bacillus subtilis* spores possess a dark repair system (spore repair) which apparently restores the photoproduct in situ to the two monomers.

3.3.4 DNA-protein crosslinks

At fluences beyond $10 \, J \cdot m^{-2}$, decreasing amounts of DNA can be extracted free of protein from irradiated cells. This effect, which presumably represents crosslinking between DNA and protein, is measurable at $23°C$, but is greatly favored at subfreezing temperatures. Resemblance of the conditions favoring DNA-protein crosslinking to those favoring spore photoproduct formation suggests similarities in the reactions involved. The importance of DNA-protein crosslinks may pertain to specific conditions, for example, where relatively few pyrimidine dimers are formed, or where they are so efficiently repaired that much rarer photoproducts become significant. Most likely, DNA-protein crosslinks would interfere with DNA replication and cause lethality, unless they are abolished by cellular repair mechanisms.

Quite a number of the common amino acids are capable of photochemical binding with nucleic acids. Cysteine is most reactive; a particular photoproduct, 5-S-cysteine-5,6-dihydrothymine (see Figure 3.8), has been implicated in DNA-protein crosslinking. An extensive review of this matter was published in 1976 by K. C. Smith.

3.3.5 Other UV photoproducts of possible biological significance

Pyrimidine hydrates. Pyrimidine hydrates were the first UV photoproducts of nucleic acid components characterized in chemical terms. The uracil product is 6-hydroxy,5-hydrouracil, which is stable to heat, while the corresponding cytosine photoproduct, 6-hydroxy,5-hydrocytosine is unstable even at room temperature. Hydration leads to loss of the 5,6-double bond (see Figure 3.8), causing lack of absorption in the 240–300 nm region.

Although pyrimidine hydrates occur commonly, their biological significance remains to be clarified. The absence of uracil in DNA and the lack of stability of the cytosine photohydrate speak against their involvement in the most frequent biological effects. However, cytosine hydrate, whose half-life under conditions prevailing in biological experiments is of the order of an hour, could possibly contribute to less frequent UV phenomena, such as mutagenesis.

Strand breakage. UV induction of single-strand breaks in DNA is relatively rare compared to the formation of previously discussed photoproducts. Nevertheless, because very sensitive assay methods are available, strand breakage can be unambiguously demonstrated and quantitatively assessed. In super-twisted, circular double-stranded DNA molecules (e.g., the replicative form of $\Phi X 174$ phage DNA) one single-strand break is easily recognized by slower sedimentation in a neutral sucrose gradient, which reflects relaxation of the super-twisted conformation. In noncircular, native DNA each break reduces the length of the respective strand, thus causing slower sedimentation in *alkaline* sucrose gradients.

Approximately one single-strand break per 10^9 daltons of DNA is produced by 254-nm irradiation at fluences from 15 to 50 J $\cdot$ m^{-2}. This is about 1/300 to 1/1000 the rate of pyrimidine dimer formation in *E. coli* DNA. The nature of these strand interruptions is not known; they could be due simply to hydrolytic splitting of phosphodiester bonds, or they could involve the sugar moiety. The first is a common intermediate step in repair of DNA and probably a normal event in the replication and recombination of an undamaged genome, and would therefore be of little consequence. The latter, however, is more likely to be lethal and could be significant even at the low rate of occurrence.

Interstrand crosslinks. The formation of crosslinks between the complementary strands of native DNA is as rare as the occurrence of strand breaks. Crosslinks would most likely interfere with replication, unless removed. Excision repair processes (see Section 7.4) abolish crosslinks caused by bifunctional alkylating agents, or by psoralen in the presence of light (see Section 13.2.2), but whether the same is true for crosslinks resulting from UV radiation is difficult to determine because of their low frequency.

The chemical nature of UV-induced crosslinks is not known. Crosslinks are probably neither "anti"-isomers of cyclobutane-type pyrimidine dimers, nor adducts nor spore photoproducts. Some degree of motional freedom of the bases seems to be required for this photochemical reaction because crosslinking fails to occur in DNA irradiated at the temperature of liquid nitrogen. Upon DNA denaturation, the crosslinks still hold together the complementary strands, permitting rapid renaturation even at low DNA concentration. In fact, this property is an excellent basis for demonstrating the existence of crosslinks.

4 Inactivation of cells and viruses

4.1 General

The most dramatic effect of UV radiation on cells and viruses is *inactivation*, that is, loss of their ability to reproduce themselves. In populations of unicellular organisms like bacteria, yeast, and others, exposed to increasing UV fluences, inactivation is experimentally recognized by a decreasing percentage of individuals capable of forming a macroscopically visible colony. Similarly, a fraction of individual cells in irradiated cultures from multicellular organisms will fail to undergo cell division, and irradiated phage or other virus particles no longer form a plaque (i.e., a macroscopically visible area of cell destruction resulting from virus multiplication) on a suitable layer of host cells. These effects are often called killing, but the term inactivation should be preferred because such nonreproducing cells or viruses sometimes retain other important properties.

Inactivation is conveniently characterized by *survival curves*, which represent the fraction of noninactivated individuals (survivors) as a function of the UV fluence. In order to discuss specific examples in this chapter and the following one, we must first understand the quantitative basis for the analysis of such curves. In later chapters we will realize that the characteristics of survival curves for a given organism depend not only on specific genetic and physiological properties, but also on outside experimental parameters, sometimes to a tremendous extent.

The foundation for understanding the relationship between the applied quantity of radiation and the observed biological effects was laid more than half a century ago by Dessauer, Blau and Altenburger, Crowther, and others. Their concept, and the formalism derived from it, has become widely known as the *target theory*. In its general form, this theory still provides a very useful basis for the quantitative analysis of radiation effects, although specific assumptions and interpretations became obsolete with the expanding knowledge in molecular biology.

The basis for this pioneering work was the belief that some peculiar *quantitative* responses of biological material to radiation can be accounted for by the quantized nature of radiation energy and the resulting statistical variation in the incidence of microphysical events. An important assumption is that, for radiation quantities typically applied in biological experiments, the mean number of these events (called *hits*) within a sensitive volume (called *target*) is relatively small. An individual cell or virus may consist of one or more targets and, as a result of random distribution of hits, any target may have received 0, 1, 2, 3, . . . hits. *Amplification processes* characteristic of living matter translate such a distribution into the observed effects; survival of an individual requires that at least one target contains less than the minimum number of hits necessary for inactivation.

One remarkable difference exists between the original concept and the current concepts, which developed from decades of experimental work. Originally, the fate of an individual was implied to be entirely determined by the primary microphysical event, but it turned out that *postirradiation conditions* have considerable impact on the studied effect. The biological net result, inactivation of a fraction of the irradiated individuals, generally reflects an intricate interrelation between the formation of lethal photoproducts and their removal by *repair processes* or similar means to avoid lethality. Thus present-day interpretations are generally more complex than originally envisaged – a familiar phenomenon in experimental science.

We should notice that the target theory was not meant to explain the biological effects of UV radiation, but rather those of ionizing radiations. The primary physicochemical effects of these radiations are strikingly different, but the analytical approaches toward understanding the observed phenomena are nevertheless similar. In both cases radiation energy causes, in crucial macromolecular structures, alterations interfering with essential biological functions. The amount of such changes is only statistically determined; therefore, the resulting biological effects can only be recorded as a function of the *mean* number of UV photons absorbed by the crucial components, or of the *average* energy of ionizing radiation delivered to them. Because these quantities can be measured, the formal target theory is appropriate for analysis and interpretation in both cases. Some of the basic ideas of the theory will be applied in the following sections to discuss examples of actually observed survival curves. Quantitative evaluations of many other UV effects employ the same or a similar approach. The reader interested in a more comprehensive treatment of the target theory is referred to books by Lea (1946), Zimmer (1961), or Elkind and Whitmore (1967).

4.2 One-hit, one-target survival curves

4.2.1 Theoretical

The simplest type of fluence-survival relationship is established if every fluence increment dF decreases the number of surviving biological units S within a population by a constant factor. We denote this by the differential equation

$$\frac{dS}{dF} = -cS \tag{4.1}$$

where the constant c characterizes the UV sensitivity of the biological unit under a given set of experimental conditions. By integration over the fluence one obtains the simple exponential function

$$S/S_0 = e^{-cF} \tag{4.1a}$$

or

$$\ln(S/S_0) = -cF \tag{4.1b}$$

where S_0 is the number of survivors at zero fluence (i.e., the initial number of viable units).

Such a relation is called a *one-hit function* or, more accurately, a one-hit, one-target function.[1] It occurs if the following conditions hold: (1) a single harmful event (hit) is sufficient to inactivate a biological unit (e.g., a cell); (2) the number of hits produced in the sensitive volume is directly proportional to the fluence applied, and their distribution among the individuals is random; and (3) the population is homogeneous with respect to radiation sensitivity of the biological units. Under such conditions, the *average* number of hits per biological unit caused by a fluence F can be expressed by cF, and, according to the Poisson distribution,[2]

$$p(n) = \frac{(cF)^n \cdot e^{-cF}}{n!} \tag{4.2}$$

the fraction $p(n = 0)$ that receives no hits equals e^{-cF}. This fraction represents the survivors, in agreement with equation (4.1a).

If the net survival is the result of primary lesions *and* postirradiation repair processes, one would expect such a simple fluence-survival relationship only if the *fraction* of primary lesions repaired is constant at all fluences. Otherwise, as is often the case, a more complex relationship applies.

According to equation (4.1b), a one-hit, one-target survival function assumes the form of a straight line with the slope $-c$ if the *natural logarithm* of the surviving fraction S/S_0 is plotted versus the UV fluence F. Therefore, it is customary (and most convenient) to use semilog paper with logarithmically subdivided ordinate, where the numerical values of S/S_0 occupy the same relative positions that the corresponding $\ln(S/S_0)$ values would on an arithmetic ordinate. The unit is the distance between 1.0 and e^{-1} (or $0.368\ldots$) on the logarithmic ordinate.

The fluence resulting in e^{-1} (or 36.8%) survival is called the *mean lethal fluence*.[3] At this fluence, cF (which expresses the mean number of hits per biological unit, *averaged over the whole population*) equals 1.0; therefore, the constant c is the reciprocal of the mean lethal fluence, including its dimensions (for example, $[m^2 \cdot J^{-1}]$). If the fluence is expressed by the number of photons per unit area, F_p, rather than by the energy per unit area, the constant c becomes σ_i, the cross section for inactivation. Correspondingly, the mean number of hits per individual at any given fluence equals $\sigma_i F_p$.

In UV photobiology a hit is considered the occurrence of a harmful photo-product that is neither successfully removed nor otherwise bypassed by cellular recovery processes. The *initial* number of these photoproducts equals the number of photons absorbed multiplied by the quantum yield for the photochemical reaction. According to equation (2.8), for irradiation under transparent conditions this would equal $\Phi\sigma F_p$. Therefore, *in the absence of any repair* we can expect from equation (2.3b)

$$\sigma_i = \Phi\sigma = 2.3 \ \Phi\epsilon q \tag{4.3}$$

or

$$\sigma_i[\mathrm{cm}^2] = 2.3 \times 10^3 \ \Phi\epsilon[\mathrm{liter} \cdot \mathrm{mole}^{-1} \cdot \mathrm{cm}^{-1}] q \ [\mathrm{mole}] \tag{4.3a}$$

However, because repair processes commonly occur, the quantum yield for inactivation, Φ_i, calculated from an experimentally measured σ_i, is usually considerably smaller than the quantum yield Φ for the underlying photochemical reaction. If f is the fraction of potentially inactivating photoproducts rendered ineffective by repair,

$$\sigma_i = \Phi_i\sigma = 2.3 \times 10^3 \ \Phi(1 - f) \ \epsilon q \tag{4.4}$$

provided that σ_i, ϵ, and q are expressed in the same dimensions as in equation (4.3a). When we compare inactivations involving damage to only one type of nucleic acid (for example, double-stranded DNA) with approximately the same base composition, both Φ and ϵ should be approximately the same in all cases. Consequently, observed differences in σ_i would be attributed to differences in the product $q(1 - f)$, and differences in the ratio σ_i/q to $(1 - f)$. This latter ratio is called the *intrinsic sensitivity*, which is of course small if a large fraction of photoproducts is repaired. Intrinsic sensitivities are mostly expressed in relative terms.

4.2.2 Examples

Although experimental data are often evaluated in terms of single-hit-single-target kinetics, and data points are approximated on a semilog plot by a straight line, in reality such kinetics are infrequently observed. Drawing a straight line is often a convenience when the experimental data are too few or show much scattering, or when the *approximate* slope of a curve is the main concern.

Current knowledge suggests that exact one-hit-one-target UV survival curves are typical only of those viruses whose sensitive material is single-stranded DNA or single-stranded RNA; examples are shown in Figures 4.1 and 4.2. Otherwise, such curves are found for the survival of particular gene functions, rather than whole individuals. In contrast, survival curves of biological systems with double-stranded DNA as the target display at least minor shoulders at low fluences, even though the remainder of the curve may be straight on a

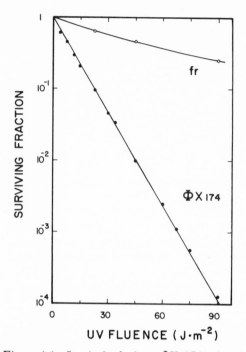

Figure 4.1. Survival of phage ΦX 174, plated on *E. coli* C, as a function of the UV fluence (254 nm). Survival of the RNA phage fr (open circles) is shown for comparison.

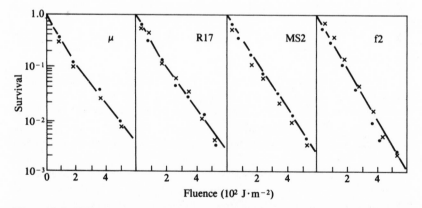

Figure 4.2. UV survival curves of different RNA phages. Plating on either wildtype cells (•) or excision-repair deficient cells (×) gives identical results. (From I. E. Mattern, M. P. van Winden, and A. Rörsch, *Mutation Res. 2*, 111, 1965.)

semilog plot. Such minor shoulders may not be detected if the quality of the experimental work is poor. Presumably the double-strandedness of the DNA constituting the usual genetic material of organisms is not compatible with true one-hit kinetics of inactivation.

An *approximately* straight line on a semilog plot can also result from fortuitous interaction of several factors, each of them causing a deviation from straight line but compensating each other. For descriptive purposes this is called a *pseudo-one-hit curve*.

4.3 Shouldered survival curves

4.3.1 Theoretical

If inactivation of an individual requires more than one hit, the effectiveness per unit UV fluence is expected to increase with fluence until it reaches some definitive value, for the following reason. At the beginning, the probability for an individual's having already received n hits (where $n > 1$) is rather small; as the irradiation continues, an increasing fraction of the survivors has already accumulated $n - 1$ hits and requires only one additional hit for inactivation. Thus the survival curve becomes gradually steeper and finally approaches the slope of the corresponding single-hit curve. Theory distinguishes the *multi-target* case, where an organism contains n distinct (usually identical) targets, each of which must receive at least one hit to become inactivated, from the *multihit* case, where a single target must be hit at least n times. In addition, combinations of both situations can be envisaged and calculated.

Multitarget case. Because from equation (4.1a) the probability for one target remaining unhit is e^{-cF}, the probability for one target being hit equals $1 - e^{-cF}$, and for n identical targets all being hit equals $(1 - e^{-cF})^n$. Thus the n-target survival function, expressing the fraction of individuals in which not all n targets are hit, is

$$S/S_0 = 1 - (1 - e^{-cF})^n \qquad (4.5)$$

which, at sufficiently low survival levels, can be approximated (using the binomial theorem) by

$$S/S_0 \approx n e^{-cF} \qquad (4.5a)$$

The latter function gives, on a semilogarithmic plot, a straight line of slope $-c$, which, as in the one-hit case, reflects the probability for a single hit. Extrapolation of this straight-line segment to zero fluence intercepts the logarithmically subdivided ordinate at n, the target number (Figure 4.3).

Multihit case. The surviving fraction in an n-hit inactivation process is the sum of the population fractions that received $i = 0, 1, 2, \ldots (n - 1)$ hits,

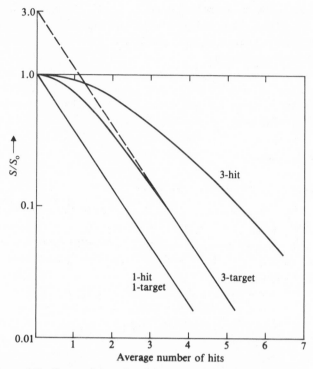

Figure 4.3. Comparison of a three-target with a three-hit curve for same target size. Extrapolation (dashed line) of the three-target curve intercepts the ordinate axis at 3.0. The one-hit, one-target curve is shown for comparison.

as calculated from the Poisson formula. If the average number of hits within this single target is cF, the n-hit survival function is expressed by

$$S/S_0 = e^{-cF} \sum_{0 \ i}^{n-1} \frac{(cF)^i}{i!} \qquad (4.6)$$

In contrast to the multitarget curve, the multihit curve becomes continuously steeper with increasing fluence, and a constant slope (of theoretically $-c$) is not reached under any practical conditions. In the theoretical example of Figure 4.3 the final slope for the three-hit curve should be the same of that for the three-target curve; it is obvious that this is far from being the case at a survival level of 10^{-2}. Thus, the n-hit survival curve, which at high fluences (where $cF \gg n - 1$) can be approximated by

$$S/S_0 \approx \frac{(cF)^{n-1}}{(n-1)!} e^{-cF} \qquad (4.6a)$$

does not lend itself to extrapolation of any well-defined straight line back to zero fluence.

Other types of shouldered curves. Besides the pure multitarget and multihit curves, other shouldered curves can arise through modifications of single-hit, multihit, or multitarget curves, particularly when repair systems are more effective at low than at high UV fluences (see Section 4.5).

General description of shouldered curves. Any shouldered curve that becomes exponential (i.e., a straight line on a semilog plot) within the experimental range of UV fluences can be partially characterized in terms of (1) the *mean lethal fluence* ($F_{0.37}$) for the straight line part, defined as the fluence increment required to decrease survival by a factor of e^{-1} (or 0.368); (2) the *extrapolation number* ($\tilde{n}$), i.e. the antilog at which the straight line extrapolation of the exponential portion intercepts the logarithmic ordinate axis; and (3) the *threshold fluence* (F_t), defined as the fluence at which the straight-

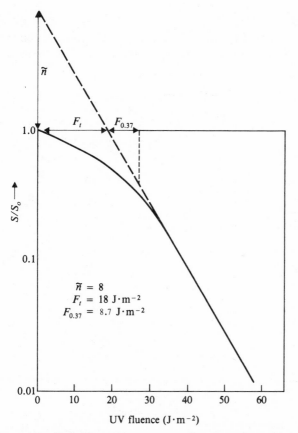

Figure 4.4. General characterization of a shouldered curve by the extrapolation number $\tilde{n}$, the threshold fluence F_t, and the mean lethal fluence $F_{0.37}$. The values shown in the panel are those describing this particular curve.

line extrapolation of the exponential portion reaches the 100 percent survival level. This is illustrated in Figure 4.4.

Because, from the definition of F_t and $F_{0.37}$, the *exponential portion of any shouldered survival curve* can be represented by

$$S/S_0 = e^{-(F-F_t)/F_{0.37}} \tag{4.7}$$

extrapolation of this portion to zero fluence gives $\tilde{n} = e^{F_t/F_{0.37}}$. Thus if any two of the quantities $F_{0.37}$, $\tilde{n}$, or F_t are specified, the third is determined. The quantity $\tilde{n}$ equals the target multiplicity n in a multitarget-single-hit case, but has no such direct meaning with other shouldered survival functions. The quantity F_t expresses in any case the approximate magnitude of fluence with which a biological system can successfully cope, irrespective of the mechanism.

4.3.2 Examples

Shouldered curves are very often experimentally observed; some typical examples are shown in Figures 4.5 to 4.8. *Multitarget*-type curves can a priori be expected in cases where suspended cells naturally exist in small groups (i.e.,

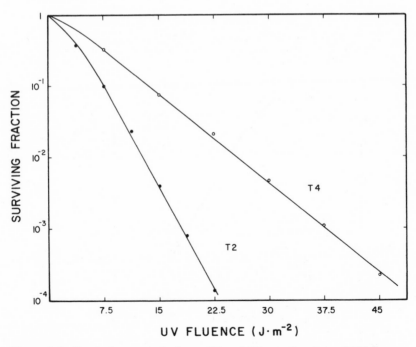

Figure 4.5. Survival curves of phages T2 and T4 as a function of UV fluence at 254 nm.

where all individuals must be inactivated to achieve loss of colony formation or where cells contain multiple sets of their genetic material). In contrast, it is difficult to picture plausible mechanisms for precise *multihit* inactivation; multihitlike curves probably reflect secondary modifications by repair, as discussed in Section 4.5.

Figure 4.5 shows survival curves of phage T2 and T4, which differ essentially by lack of a repair mechanism in T2 (see Section 7.5). The slight shoulder is not only characteristic of the T-even phages, but also of the T-odd and many other phages of *E. coli*, although in T1, T3, and T7 it is sometimes overlooked because of the compensating upward concavity of these curves at higher fluences (see Section 4.4.2). In view of the absence of shouldered curves for phages containing single-stranded DNA or RNA, double-strandedness of DNA is the likely reason for these minor shoulders. Accurate determination of the low-fluence portion reveals a slope compatible not with simple two-target kinetics, but rather with a combination of one-target and two-target kinetics.

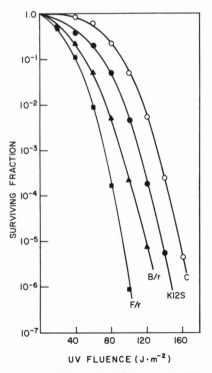

Figure 4.6. Different extents of shoulders in the survival curves of the *E. coli* wildtype strains C, K12S, B/r, and F/r, irradiated at 254 nm under identical experimental conditions. (From W. Harm and K. Haefner, *Photochem. Photobiol. 8*, 179, 1968.)

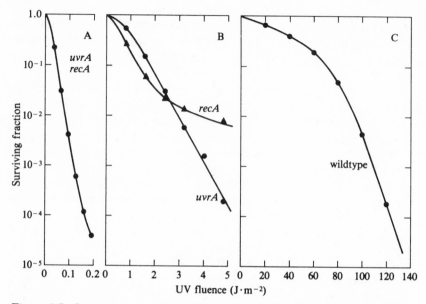

Figure 4.7. Survival curves of stationary phase cells of various *E. coli* K12 derivatives. *Panel A*: Strain AB 2480 (*uvrA recA*), which is the most UV-sensitive *E. coli* strain in existence (presumably because of complete lack of dark repair). *Panel B*: Strains AB 2437 (*uvrA*) and AB 2463 (*recA*), deficient in excision repair or Rec-repair, respectively. (The two markers are identical with those combined in strain AB 2480). *Panel C*: Wildtype K12 cells for comparison. Relative to the fluence scale on panel A, the scale in panel B is condensed 10-fold, and that in panel C 200-fold.

Large shoulders are typical of survival curves of many bacteria and yeasts. Examples of various wildtype *E. coli* strains are shown in Figure 4.6. In general, such curves are more prone to show variations from one experiment to another than those of viruses, particularly if not all experimental parameters are strictly controlled. The reason is extensive repair, as even small modifications in its extent may considerably alter the shoulder portion of the curve, and thus the total survival. The general shape of curves like those in Figure 4.6 most likely expresses at increasing UV fluence a progressive reduction of the percentage of damage repaired. This interpretation is consistent with the fact that survival curves of repair-deficient strains have only slight shoulders, compared with wildtype. Figure 4.7 shows examples for mutants of *E. coli* K12 deficient in either excision repair, postreplication repair, or both types of repair.

An extremely large shoulder is characteristic of *Micrococcus radiodurans*, an unusually radiation-resistant bacterium, as illustrated in Figure 4.8. Its survival remains close to the 100 percent level up to about 400 J $\cdot$ m^{-2} of 265-nm radiation, a fluence that would leave no single survivor among 10^8

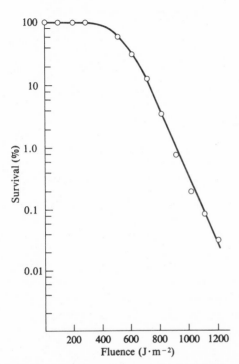

Figure 4.8. Survival of wildtype cells of *Micrococcus radiodurans* in log phase as a function of 265-nm UV fluence. (From J. K. Setlow and D. E. Duggan, *Biochim. Biophys. Acta 87*, 664, 1964.)

cells of the most resistant *E. coli* strains. This bacterium is also extremely resistant to ionizing radiations and decay of radioactive phosphorus. Its UV survival kinetics resembles a 10^3-target curve, but interpretation of the shape of the curve in terms of the target theory would be unrealistic. It is likely that the repair systems in this organism are so effective that up to a certain fluence virtually every potentially lethal lesion is repaired or its consequence bypassed.

4.4 Upward concave survival curves

4.4.1 Theoretical

Two-component curves. If a population consists of two fractions of individuals, f_1 and f_2, being inactivated with one-hit kinetics but at different rates, the type of fluence survival curve shown in Figure 4.9 will result. It is expressed by

$$S/S_0 = f_1 e^{-c_1 F} + f_2 e^{-c_2 F} \tag{4.8}$$

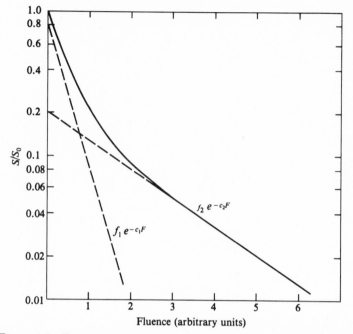

Figure 4.9. Two-component survival curve (solid line) of the type $S/S_0 = f_1 e^{-c_1 F} + f_2 e^{-c_2 F}$. The single components are represented by the dashed lines, where $f_1 = 0.8$, $f_2 = 0.2$, and $c_1 = 5c_2$.

At sufficiently high fluences the decrease in survival is essentially determined by one of the two terms of the equation, provided c_1 and c_2 differ considerably. If $c_2 < c_1$, extrapolation of the straight line segment of the survival curve toward the ordinate axis corresponds to the function $f_2 e^{-c_2 F}$, and by subtracting the values for this component from the total survival one obtains $f_1 e^{-c_1 F}$. Obviously, these partial survival curves have their origins on the ordinate axis at points f_1 and f_2, and have slopes of $-c_1$ and $-c_2$, respectively.

Multicomponent curves. The same procedure resolves, in principle, a curve consisting of more than two components, but each successive subtraction increases the degree of uncertainty. For the extreme case of a continuous distribution of sensitivities, with individuals of each sensitivity class being inactivated according to one-hit kinetics, the sum in equation (4.8) is replaced by an integral:

$$S/S_0 = \int_{c_{\min}}^{c_{\max}} w(c) e^{-cF} \, dc \tag{4.9}$$

Here $w(c)\ dc$ is the fraction of the population whose sensitivity is expressed by a value between c and $c + dc$. Curves of this type, if plotted as $\log S/S_0$ versus fluence, are characterized by a continuously decreasing slope. They are found for UV inactivation of bacterial transforming DNA (see Figure 5.1A) and will be discussed in more detail in Section 5.1.

4.4.2 Examples

Examples for two-component UV survival curves are obtained with phages T1 and T3 (Figure 4.10). Here, one fraction of the phage population undergoes extensive host-cell reactivation, and therefore appears to be less sensitive than the other fraction, in which host-cell reactivation is less apparent. A phage population may thus be homogeneous regarding its primary UV sensitivity, but become heterogeneous in the course of postirradiation processes. Similar examples are the coliphage T7 and the *Haemophilus* phage HP1. Extrapolation of the shallower portions of the survival curves in Figure 4.10 to zero fluence indicates that approximately 30 percent of the T1 phage, and 10

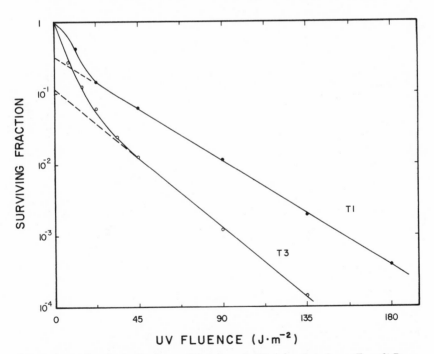

Figure 4.10. Survival of phages T1 (•) and T3 (○), plated on *E. coli* B, as a function of the 254-nm UV fluence. The dashed lines are extrapolations of the lower curve portions, indicating the approximate fraction of individuals undergoing maximal host-cell reactivation.

percent of T3 phage recover maximally owing to host cell repair. At least in the T1 example, one recognizes a shoulder in the curve portion corresponding to the more UV-sensitive fraction of the population.

Among bacteria, upward-concave survival curves are well known for the *E. coli* strain B and some of its derivatives (for example, B_{s-1}) and *recA* mutants of *E. coli* K12. Usually the steep initial part of the curve represents the great majority of the cells; the less sensitive component sometimes displays a downward trend, making the whole curve sigmoidal (see Figure 4.11). The exact shape of such curves depends greatly on the particular experimental conditions. Differences in sensitivity of individuals reflected by the shape of the curve are usually nonhereditary, which is easily demonstrated by testing survivor colonies from different portions of the curve for their UV sensitivi-

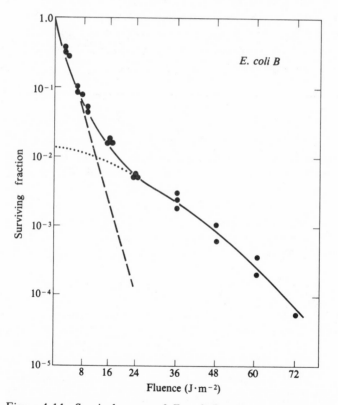

Figure 4.11. Survival curve of *E. coli* B cells approaching stationary phase. The slope characteristic of the great majority of individuals is represented by the dashed line; extrapolation of the lower portion of the curve to zero fluence (dotted line) indicates that roughly 1% of the population are much less sensitive, resembling in this respect B/r cells. (From W. Harm, *Mutation Research 4*, 93, 1967; modified.)

ties. Except for the occurrence of rare mutants, cultures of such survivor colonies display essentially the same survival curve as the parental culture.

4.5 Modification of survival kinetics

Of necessity, analysis of survival kinetics relates only to those radiation-induced alterations that eventually cause inactivation. These are the photoproducts remaining effective after the cells have exhausted their repair potentials or other means to bypass damage. Such processes modifying the net result of a radiation treatment are primarily determined by the genotype of the biological system, but their extent is often affected by specific experimental conditions before, during, or after irradiation.

Demonstration of the basic (i.e., unmodified) UV sensitivity of a biological system requires the availability of mutants with an absolute lack of repair. Otherwise, conditions would have to be applied to render existing repair systems completely nonfunctional, a situation difficult to achieve. Comparison between mutants with absolute lack of repair and those having only one repair system intact establishes the extent of this particular type of repair, and application of several different pre- or postirradiation conditions reveals their modifying effects on the repair. Unless the effects of repair or other modifications affecting the survival are studied under these rigid rules, an accurate quantitative assessment is not possible. Unfortunately, this difficulty is often met because completely repairless systems are only rarely available (for example, in *E. coli* K12 and perhaps in yeast). In wildtype cells usually several repair systems are engaged in removal of the same types of lesions, so that the contribution of each one of them is difficult to determine. Nevertheless, even under less favorable conditions it is possible to determine, in one way or another, the approximate magnitude of modifying effects by repair. Extensive experimental work has established the following general types of modification patterns.

(a) **Constant fluence modification.** If a process always eliminates a constant fraction of potentially lethal photoproducts, it is as if each fluence had been reduced by a proportionate amount. In such a case, the general shape of the curve would remain unaltered, but the fluence required for any given effect would be higher by a constant factor. The ratio between the two fluences producing identical effects is called the *fluence reduction factor* or *fluence modification factor*. It is expressed either by the ratio of the smaller over the larger fluence or by the reciprocal of this ratio; fortunately, this inconsistency does not lead to confusion, for in the former case the value is always <1, and in the latter case >1. A clear-cut example of this modification pattern in survival is *photoenzymatic repair*, where a major fraction of lethal UV photoproducts is abolished by an enzyme in the presence of light (see Section 7.2).

(b) **Declining fluence modification with increasing fluence.** If a repair system functions efficiently at low radiation fluences, but at higher fluences becomes either "overburdened" or damaged itself, the fluence modification factor should gradually decrease. The resulting tendency of survival curves to become steeper with fluence increase is the most likely reason for observing survival curves with large shoulders (see Section 4.3.2).

(c) **Population heterogeneity with respect to fluence modification.** A population may be phenotypically or genotypically heterogeneous regarding some modifying process. If one fraction of the population is extensively modified, while another fraction is little or not at all modified, two- (or multi-) component curves resembling those in Figure 4.10 will result.

(d) **Combination of several fluence modification effects.** While any population heterogeneity in the radiation response tends to make the survival curve upwardly concave, a decrease in the extent of repair at high fluence tends to make the curve upwardly convex. Thus the combination of these effects can result in curves of various sigmoidal shapes.

An example, shown in Figure 4.11, is strain B of *E. coli*, whose UV survival is highly dependent on postirradiation parameters. Compensation between the two opposing tendencies within a wide fluence range may result in the approximation of a straight line (pseudo-single-hit curve).

Fluence modification by the occurrence of repair processes or by alteration in the extent of repair processes will be discussed in detail in Chapters 7 and 8.

4.6 Intrinsic sensitivities

Equation (4.4) showed that for one-hit survival kinetics the UV sensitivity of individuals depends, under otherwise similar conditions, on the quantity q of the sensitive material and its extinction coefficient ϵ, on the quantum yield Φ for the photochemical reaction underlying inactivation, and on the extent of modification (by repair or similar processes) expressed by f. From the measured sensitivity (expressed by σ_i) one can calculate the *intrinsic sensitivity* (σ_i/q), by dividing by the amount of the UV-sensitive material. Differences in σ_i/q values for different biological systems must thus be related to either ϵ, Φ, or f.

Table 4.1 shows examples for various double-stranded DNA phages. Compared with T5 (whose sensitivity is arbitrarily set equal to 1.0), other phages have considerably lower intrinsic sensitivities, although all of them contain the four usual nucleotide bases in similar proportions so that their ϵ and Φ values should not much differ.[4] In these cases, the lower intrinsic sensitivities (relative to T5) express lower $(1 - f)$ values resulting from repair in the host cell (host cell reactivation; see Section 7.3). The extent of repair determined

Table 4.1 *Relative intrinsic UV-sensitivities of various phages with double-stranded DNA*

Phage	Survival range considered	Mean lethal fluence $(J \cdot m^{-2})$	$F_{0.37}$ relative to that of T5	Number of nucleotide pairs	Relative intrinsic UV sensitivity
T5	below 10^{-1}	3.25	1	1.3×10^5	$\equiv 1.0$
T3	below 10^{-2}	20	6.15	3.1×10^4	0.68
T1	below 10^{-1}	26.8	8.25	5.0×10^4	0.31
λ	below 10^{-1}	36	11.1	5.3×10^4	0.22

independently by comparison of phage survival in repair-proficient and repair-deficient strains is greatest for λ, smaller for T1, and still smaller for T3, which is consistent with the relative intrinsic sensitivities shown in the table.

Differences in intrinsic sensitivities of materials less closely related than those compared in Table 4.1 may not be simply attributed to repair. If the type and base composition of nucleic acid is known, ϵ and Φ may be estimated or actually measured, so that eventually remaining differences can be attributed to f. Conversely, if the extent of repair or possible other factors determining f are known, relative values for Φ and/or ϵ can be estimated. For example, comparison of Figures 4.1 and 4.2 indicates that the RNA phages have only about 1/10 the intrinsic sensitivity of the single-stranded DNA phages ΦX 174 and others, which contain about the same amount of nucleic acid. As far as known, both types of small phages undergo no appreciable repair. Because ϵ is similar for both types of nucleic acids, the large difference in intrinsic sensitivities is likely to reflect a corresponding difference in the quantum yield for the formation of lethal photoproducts.

5 Inactivation of genetic material in vitro

5.1 Inactivation of bacterial transforming DNA

In bacterial transformation, DNA from one bacterial strain (donor strain) is taken up through the medium by cells of another bacterial strain (recipient strain), which differs from the first one in at least one genetic character. Incorporation of donor DNA into the recipient chromosome alters the genotype of the cell, that is, the cell becomes transformed. The frequency of transformation events depends on such factors as the fraction of competent cells in the recipient population (i.e., cells capable of taking up DNA), the concentration of the DNA solution, the genetic marker for which transformation is tested, and others. However, for a given set of conditions the observed frequency of cells undergoing transformation is well reproducible.

Transformation experiments make it feasible to treat DNA in vitro with physical or chemical agents and to study the biological consequences of such treatments in the recipient cells. Studies involving UV irradiation of transforming DNA and its repair in vitro have been carried out mainly with *Haemophilus influenzae*, and to a lesser extent with *Bacillus subtilis* or *Diplococcus pneumoniae*. In *Haemophilus*, in contrast to the latter two species, virtually all cells can be made competent, so that a relatively large number of transformed cells can be obtained. Therefore, the following description of results refers to *Haemophilus influenzae* DNA only, but is likely to apply also to other kinds of transforming DNA.

When transforming DNA is exposed to UV radiation prior to its addition to competent cells, the transforming activity, expressed by the fraction of transformed recipient cells, decreases. If by T_0 and T we denote the number of transformed cells obtained with unirradiated and irradiated DNA, respectively, the ratio T/T_0 defines the survival of transforming activity. For a *single genetic marker*, such as high level resistance to streptomycin (1000 μg/ml), a semilogarithmic plot of T/T_0 versus the UV fluence displays an upwardly concave curve, as shown in Figure 5.1A. This curve is fairly well described by the function

$$T/T_0 = 1/(1 + cF)^2 \tag{5.1}$$

where c is a marker-characteristic constant, and F is the UV fluence. Equation (5.1) can likewise be written in the form

$$1/(T/T_0)^{1/2} = (T_0/T)^{1/2} = 1 + cF \tag{5.1a}$$

58

Therefore, if experimental data are plotted in the form of $(T_0/T)^{1/2}$ versus F, they will give a straight line with slope c, intercepting the ordinate axis at 1. Figure 5.1B shows the experimental data points from panel A plotted in this manner; deviations from a straight line are minor.

The constant c characterizing the relative UV sensitivity of transforming activity can vary considerably from one marker to another. For example, c is 5–10 times higher for the high level streptomycin marker than for the markers Nb_1 (low level novobiocin resistance) or E_1 (erythromycin resistance). This great difference reflects mainly a different extent of excision-resynthesis repair of the UV lesions in DNA (see Section 7.4). Accordingly, under conditions of repair inhibition or in an excision-repair-deficient recipient cell the sensitivities of the markers become similar.

UV survival of transforming activity for *two loosely linked markers* (i.e., markers that can be transferred on the same DNA fragment, but are not in close vicinity), or for *two completely unlinked markers* entering the same cell, is expressed by the analogous inverse fourth power function:

$$T/T_0 = 1/(1 + cF)^4 \tag{5.2}$$

This function gives a straight line if $(T_0/T)^{1/4}$ is plotted versus F.

How can inactivation of transforming DNA and its quantitative dependence upon UV fluence be explained in molecular terms? Transformation involves at least three distinct steps: uptake of DNA through the cell wall, homologous pairing with the recipient chromosome, and, finally, integration into the latter. Experiments have shown that the first step is little affected by fluences

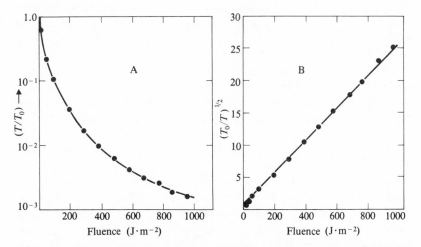

Figure 5.1. Survival of high-level streptomycin resistance marker of *Haemophilus influenzae* transforming DNA. The same experimental data plotted in panel A as log T/T_0 vs. fluence, are plotted in panel B as $(T_0/T)^{1/2}$ vs. fluence.

up to 10^3 J $\cdot$ m^{-2}; thus inactivation of transforming activity must result from interference of DNA photoproducts with homologous pairing and/or *viable* integration of the marked DNA region.

The upward concavity of marker survival curves on a semilog plot (Figure 5.1A) tells us that the probability for marker inactivation per unit UV fluence continually decreases with increasing fluence. Formally they can be considered multicomponent-type curves, as mentioned in Section 4.4, where all components differ from one another in UV sensitivity. The possibility that these components are represented by the greatly varying lengths of marker-carrying DNA fragments has been experimentally ruled out. However, the unusual shape of the curves can be satisfactorily explained by the *marker rescue model*, to be discussed next.

Although many details of the processes involved in transformation are not known, it is evident that integration of the marked donor DNA into the recipient chromosome requires genetic recombination on each side of the marker, which in the simplest case may be just one altered nucleotide pair. Therefore, the location of UV lesions in the irradiated DNA fragment relative to the presumably random sites of recombination events will determine whether or not they result in marker inactivation. Only when the recombination sites are on both sides closer to the marker than the nearest *unrepaired* UV lesions (where the repair could occur before, as well as after integration), is the marker being "rescued" from the remainder of the damaged DNA molecule and expected to give rise to a viable transformed cell.

This model clearly predicts a marker survival curve with continually decreasing slope in semilog plot. Because only the hits (or unrepaired UV lesions) closest to the marker are relevant, the target for any additional hit is represented by the nondamaged DNA region to the left and right of the marker. This region becomes smaller as the fluence increases; therefore, any additional relevant hit requires on the average a greater UV fluence than the previous one. Or, to view matters differently, the closest recombination events on each side of the marker define the target and determine – in retrospect, because the UV lesions are already there – whether or not the target is hit. The multicomponent character of the marker survival curve then simply reflects the range of target sizes resulting from the random location of recombination events.

In quantitative terms, one equates the fraction of surviving markers with the probability that the nearest recombination event on each side of the marker will lie closer than the nearest site of unrepaired injury. For simplicity, we consider a marker of small extent (e.g., a point mutation) lying near the middle of a long DNA fragment. We assume further that sites of recombination occur at random with a mean frequency b per unit length of the DNA strand, whereas at a given UV fluence hits occur independently with a mean frequency h per unit length. Thus on either side of the marker the probability

P for a recombination event being closer to the marker than the nearest hit is $(b/b + h)$, so that the relative marker survival should be expressed by the product of the two probabilities:

$$P^2 = \frac{b^2}{(b + h)^2} = \frac{1}{[1 + (1/b)h]^2} \tag{5.3}$$

This function is identical with equation (5.1) if one equates P^2 with the fractional marker survival T/T_0, taking into account that h is proportional to the fluence.

The inverse fourth power survival function for two loosely linked markers (equation 5.2) can be explained by the same model. Except at very low UV fluence, loosely linked markers must be rescued from the damaged DNA region between them, so that both of them survive only when recombination events integrate them into the recipient DNA independently of one another. The same, of course, is expected for completely unlinked markers. In either case, the probability for survival of both markers should resemble the product of the single marker survival probabilities. These considerations strictly predict an inverse fourth power function only for markers with equal inactivation constants c, but it has been found experimentally that even marker pairs with rather different inactivation constants approximate such a function.

At present, the marker rescue model offers the best explanation for the UV inactivation kinetics of transforming DNA, and is supported by several other results. For example, transforming DNA sheared to fragments of considerably lower average molecular weight than the usual average of 10^7 to 4×10^7 daltons shows decreased UV sensitivity, as expected from the marker rescue model. Furthermore, inactivation by other agents causing localized lethal damage in DNA without destroying its integrity should follow a similar function as found for UV; this is indeed the case for nitrous acid, hydrazine, and hydroxylamine.

Thus UV studies on transforming DNA seem promising regarding our understanding of the relationship between DNA damage and genetic recombination, which is important in several respects (see Sections 7.6 to 7.8, and 10.4, 10.5, and elsewhere). There is experimental evidence that UV lesions, together with the genetic marker, can be integrated into the recipient chromosome and inactivate the recipient cell, unless they are repaired. On the other hand, results clearly show that UV lesions located on unmarked fragments cause little lethality in the recipient cell population. We thus conclude there must be strong discrimination by the recombination process against incorporation of UV-damaged DNA segments.

We saw in the previous chapter that bacterial or phage survival curves are essentially independent of the initial titer S_0 in nearly transparent suspensions. Similarly, the survival of transforming activity is essentially independent of the DNA concentration employed. Only a close look reveals a concentration

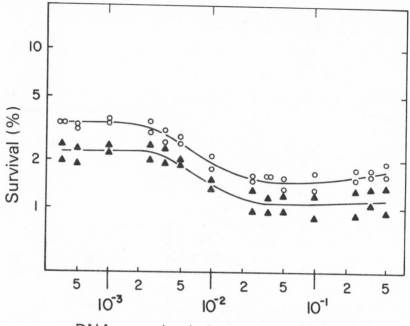

DNA conc (μg/ml of transf. mixture)

Figure 5.2. Survival of a high-level streptomycin resistance marker of *Haemophilus influenzae* as a function of DNA concentration in the transforming mixture. The DNA was UV (254 nm) irradiated either with $192 \text{ J} \cdot \text{m}^{-2}$ (○) or $240 \text{ J} \cdot \text{m}^{-2}$ (▲). (From H. Harm, *Molec. Gen. Genetics, 107,* 71, 1970.)

dependence, indicating more inactivation at high than at low DNA concentrations (Figure 5.2). Transition between the two levels occurs in a concentration range where the average uptake of DNA molecules (marked *and* unmarked) per cell is neither much above nor much below 1.0. Thus, most likely, the lower survival observed at the higher DNA concentrations is due to competition by co-infecting UV-damaged, nontransforming DNA fragments for intracellular repair.

5.2 Inactivation of viral nucleic acids

Infectivity of naked viral nucleic acids was discovered later than bacterial transformation and has been less extensively investigated in UV studies. Although such infections usually occur at rather low rates, they offer certain advantages. In contrast to variably sized and composed transforming DNA molecule fragments, infectious viral nucleic acid can be prepared essentially homogeneous. Because successful infection, resulting in production of viable progeny, requires intactness of the whole virus genome, evaluation of biological experiments refers to nucleic acid molecules of identical (full) lengths, even if the preparation also contained fragmented molecules. Viral genomes may consist of DNA or RNA, and sometimes the same genome can be compared in single-stranded and double-stranded form. Additional flexibility in experimental procedures derives from the fact that there is no need to preserve the cells as colony-forming units. These features may outweigh the usually low efficiency of viral nucleic acid in initiating successful infection.

Bacterial cells can be infected under at least three different conditions with pure bacteriophage DNA or RNA: (1) cells are converted to spheroplasts prior to infection so that the cell wall no longer acts as a barrier; (2) cells competent for transformation can be infected with DNA of phage normally infecting the same strain; and (3) cells preinfected with intact phages undergo alterations permitting subsequent invasion of naked DNA from the same or a closely related type of phage. Because of the similarities to the bacterial transformation process, virus nucleic acid infection is sometimes referred to as *transfection*.

Single-stranded DNA. DNA from phage ΦX 174 is a covalently closed, single-stranded circle of about 1.7×10^6 daltons molecular weight. Its UV inactivation is of the single-hit type at all wavelengths, the action spectrum closely resembling DNA absorbance. The sensitivity of this DNA at wavelengths >240 nm is virtually identical to that of the intact phage, indicating that the UV damage to the viral particle is entirely in DNA (see Figure 3.7). At shorter wavelengths, the free DNA incurs less lethal damage than the complete phage, possibly because additional photochemical DNA-protein interactions occur in the latter (see Section 3.2).

Comparison of single-stranded with double-stranded DNA. Like the viral single-stranded ΦX 174 DNA, the double-stranded replicative form (which is synthesized after this phage has infected a host cell) is inactivated exponentially with UV fluence. However, Figure 5.3 shows that its UV sensitivity is only about 1/10 that of the single-stranded form when assayed in *wildtype E. coli* spheroplasts. In contrast, infection of spheroplasts from excision-repair-deficient cells shows much greater sensitivity of the replicating form, namely 0.6–0.7 that of the single-stranded DNA, whereas the single-stranded DNA is equally sensitive in either type of spheroplast. This is an excellent demonstration of the requirement for double-stranded DNA in excision-resynthesis repair (see Section 7.4), which is implicit from its mechanism. The remaining difference in sensitivity between single- and double-stranded infectious DNA in the absence of repair may be explained either by the hyperchromicity of single-stranded DNA, or by quantitative or qualitative differences in the photoproducts formed.

The large difference in UV sensitivity between single- and double-stranded

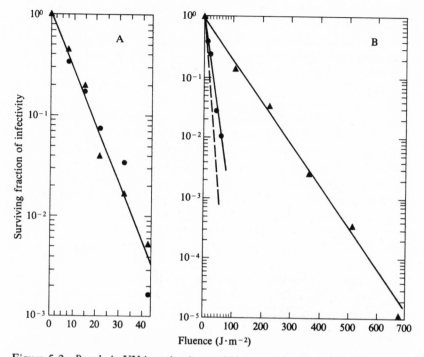

Figure 5.3. *Panel A*: UV inactivation at 258 nm of single-stranded DNA from phage ΦX 174, assayed on spheroplast of either repair-proficient (▲) or repair-deficient (●) *E. coli* cells. *Panel B*: UV inactivation at 270 nm of double-stranded DNA from phage ΦX 174, assayed on the same repair-proficient (▲) and repair-deficient (●) strains. Inactivation of the single-stranded nucleic acid under the same conditions is shown by the dashed line. (From M. Yarus and R. L. Sinsheimer, *J. Mol. Biol.* 8, 614, 1964.)

DNA assayed in repair-proficient spheroplasts has been used by Sinsheimer and colleagues for determining the proportion of these two forms within cells after infection with intact ΦX 174 phage.

Double-stranded DNA. Double-stranded DNA of phage HP1 is infectious for competent cells of *Haemophilus influenzae*. Like the UV survival curve of the intact phage, the curve of the free DNA is of the two-component type, but the sensitivity of the latter is slightly greater. The two components of the curve apparently reflect different extents of excision repair (as is typical for phages T1, T3, and others) because the low-sensitivity component is not observed in the presence of a repair inhibitor, acriflavine, or in excision-repair-deficient host cells.

Infective RNA. Single-stranded RNA molecules from *tobacco mosaic virus* can be utilized for successful infection of tobacco leaves and production of complete virus progeny. *The UV-sensitivity of the naked RNA is much greater than that of the intact virus*, as a consequence of differences in the photoproducts occurring under the two conditions. Pyrimidine dimers are produced in free RNA but not in the intact virus, where the RNA is held in a rigid position by coat protein units wrapped around it. Therefore, UV-irradiated free RNA can be photorepaired after infection (see Section 7.2), in contrast to either the irradiated intact virus or the infectious RNA extracted from it. In the latter case, UV inactivation resembles that of the intact virus, which is possibly the result of photochemical RNA-protein interactions.

Double-stranded RNA is the genetic material of some animal viruses (e.g., *Reoviruses*), which form single-stranded RNA only after infection. Conversely, other animal viruses (e.g., *poliovirus*) contain single-stranded RNA, but form double-stranded RNA after infection. Infection of HeLa cells by UV-irradiated, infectious single-stranded poliovirus RNA shows that it is only slightly more sensitive than the double-stranded form, suggesting that no excision repair occurs. Thus far, there is no evidence for any kind of dark repair of RNA irradiated either in the intact virus or in suspension.

6　Causes of lethality

6.1　General

The quantum yields for inactivation of cells or viruses by far-UV radiation are usually low, of orders of magnitude from 10^{-3} to 10^{-6}. Thus the vast majority of UV photons absorbed by nucleic acids or other molecules do *not* cause lethality. Most frequently, their energy is dissipated without leading to a photochemical reaction, and even where a photoproduct is formed, it does not necessarily result in inactivation. Figure 6.1 illustrates schematically a number of possibilities.

If a photoproduct is formed, it may be either (a) *biologically irrelevant* (i.e., the organism does not recognize the difference from the original condition), (b) *biologically relevant under all practical circumstances,* or (c) *potentially relevant.* The last case is most common: Potentially lethal photoproducts are formed, but the biological consequences remain obscure for some time, until postirradiation cellular repair processes determine the final result. If repair processes are successful in preventing lethality, they may still lead to such nonlethal effects as *mutation, growth delay,* increased frequency of *recombination,* or others.

Possibly the latter effects could also be caused by primarily nonlethal photoproducts. However, at present there is little evidence for this as well as for any of the other theoretical possibilities indicated in the scheme by dashed lines with question mark. The purpose of the scheme is merely to provide a basis for theoretical thought and for discussions further below; it should neither be considered final nor complete. Obviously the solid arrows can represent various types of photoproducts, whose biological consequences not only depend on the particular biological system, but also on the particular location of a photoproduct in the genome.

It is important to relate photochemical processes in DNA or other biomolecules with observable photobiological effects. In the case of lethality the first step is to ascertain *which* photoproducts contribute, whereas the second step is to determine the *mechanisms* by which such photoproducts interfere with the vital biological functions. In addition, even the recognition of biologically irrelevant photoproducts in biomolecules helps to improve the overall picture.

66

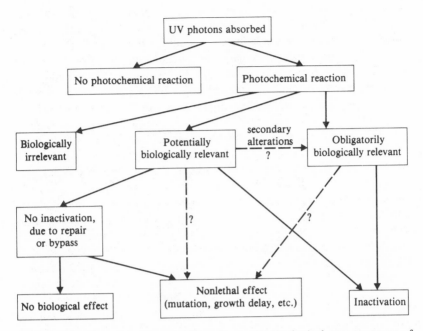

Figure 6.1. Schematic representation of possible biological consequences of UV photons absorbed in DNA (see text.)

6.2 Evidence for lethal photoproducts

6.2.1 Pyrimidine dimers

Attempts at relating known photoproducts to defined biological effects are based on common characteristics. Take, for instance, the quantum yield as a criterion. Any photochemical reaction occurring with a quantum yield considerably lower than that for inactivation can be excluded as a major cause of the lethal effect. Under conditions involving little or no repair (which admittedly are exceptions), this applies for photochemical reactions with quantum yields below 10^{-3} to 10^{-4}. To exclude photochemical reactions with higher quantum yields as a cause of lethality, other criteria must be applied. For example, irradiated bacteriophages can be stored in buffer at refrigerator temperature for days or weeks without noticeable change in the fraction surviving. Thus the underlying photoproducts cannot be of a kind that reverts under these conditions.

During the past two decades, considerable progress in this line of research has been made through the study of repair processes. For example, it was known that in many irradiated cellular or viral systems 50 to 90 percent of potentially lethal lesions can be abolished by subsequent illumination with near UV or visible light, in a process called *photoenzymatic repair* (see Section 7.2). Under the same experimental conditions, it was found that a similar or even larger fraction of cyclobutyl pyrimidine dimers is restored to monomers. The same is observed when irradiated transforming DNA, together with photoreactivating enzyme, is exposed to near UV or short wavelength visible light in vitro, whereas neither the enzyme alone, nor light alone, repairs lethal lesions or monomerizes dimers. This strong correlation is convincing evidence that pyrimidine dimers are usually the great majority of potentially lethal photoproducts.

Other evidence is furnished by photochemical rather than photoenzymatic monomerization of dimers. We saw in Figure 3.10 that the ratio of thymine dimer absorption over thymine absorption decreases by about three orders of magnitude from 230 to 285 nm. Consequently, shorter wavelengths strongly favor monomerization of existing dimers relative to dimerization of monomers in DNA, so that in the photosteady state the ratio $T\diamond T/T$ should be much smaller at 230 than at 285 nm. Because the quantum yield for monomerization of dimers in DNA is roughly 100-fold higher than for dimer formation from monomers, R. B. and J. K. Setlow made the following prediction: If pyrimidine dimers were a major cause of lethality, irradiation at 239 nm should decrease the number of dimers resulting from a previous exposure to high fluences of 280-nm radiation, and thereby *reduce* the lethal effect.

Because of the required high fluence at 280 nm, most cellular organisms and viruses are unsuited for this kind of work, but bacterial transforming DNA is sufficiently resistant (see Section 5.1) to meet the conditions. The results presented in Figure 6.2 fully agree with the prediction. Irradiation at 239 nm following exposure to high fluences of 280-nm radiation recovers transforming activity, whereas 280-nm irradiation following exposure to high fluences of 239-nm radiation further decreases transforming activity.

Further evidence that pyrimidine dimers are potentially lethal comes from *dark repair* processes. A major mechanism (excision-resynthesis repair; see Section 7.4) involves excision of pyrimidine dimers containing nucleotide sequences from irradiated double-stranded DNA, and resynthesis of the excised portion. Mutant strains of bacteria, yeast, mammalian cells, and phage T4, deficient in performing these reactions, display greatly increased UV sensitivity. Although excision repair is not dimer-specific, but applies also to several other photochemical or chemical alterations in the DNA molecule, the suggestion that cyclobutane dipyrimidines play a predominant role in lethality is strengthened by a number of other correlations. For example, presence of caffeine or acriflavine lowers the rate of dimer excision from DNA and likewise increases the net UV sensitivity of most microbial systems

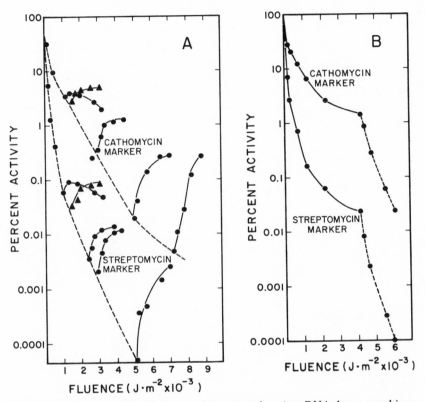

Figure 6.2. Inactivation of *Haemophilus* transforming DNA by a combination of 280-nm (dashed line) and 239-nm (solid line) radiation. *Panel A*: If 280-nm radiation is applied first, subsequent irradiation with 239 nm restores some of the activity. *Panel B*: If 239 nm is applied first, subsequent 280-nm irradiation decreases the activity further. (From R. B. Setlow and J. K. Setlow, *Proc. Natl. Acad. Sci. U.S. 48*, 1250, 1962.)

(see Section 7.4.1), whereas treatment of DNA with acridines *prior* to UV irradiation results in reduced dimer formation and enhanced survival (see Section 13.2.3).

Thus pyrimidine dimers are the overwhelming cause of lethality under conditions of complete repair deficiency, but their importance for the finally observed lethal effect decreases by the extent to which they are repaired or bypassed in some way. As a corollary, the relevance of less frequent, but unrepairable, photoproducts increases; and in cases of extremely high UV resistance (e.g., in *Micrococcus radiodurans;* see Section 4.3.2) they can become predominantly responsible for the observed inactivation.

6.2.2 Spore photoproduct

Another photoproduct that can be related to lethality is the spore photo-product: *5-thyminyl-5,6-dihydrothymine* (TDHT; see Section 3.3.3). A fluence of 100 J · m^{-2} of 265-nm radiation converts, in *Bacillus megaterium* spores, 2.5–5 X 10^4 thymine residues per genome into spore photoproducts. At this fluence most spores survive, but still higher fluences inactivate an increasing fraction of spores, as seen in Figure 6.3. These results indicate a high extent of cellular repair of spore photoproducts.

That the spore photoproduct is in fact the major cause of UV inactivation of spores is indicated by various experimental results. Increased concentrations of Mn^{2+} ions in the growth medium lead to less TDHT formation after UV irradiation and to proportionally less inactivation. Spores irradiated at either room temperature, -80°C, or -196°C convert 5, 8.4, and 1.4 percent of their thymine residues, respectively, to spore photoproduct and the corresponding survival levels of survival are 0.15, 0.008, and 65 percent. These photoproducts disappear from DNA during spore germination and reappear as acid-soluble material in the cells or the surrounding medium. In addition,

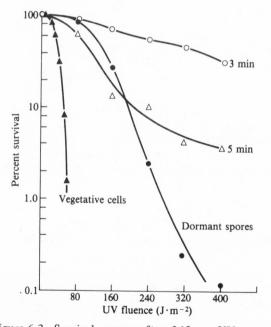

Figure 6.3. Survival curves after 265-nm UV exposure of vegetative cells, dormant spores, and germinating spores of *Bacillus megaterium*. Germinating spores were exposed at either 3 or 5 min after dilution into germination medium. (From R. S. Stafford and J. E. Donnellan, Jr., *Proc. Natl. Acad. Sci. U.S. 59*, 822, 1968.)

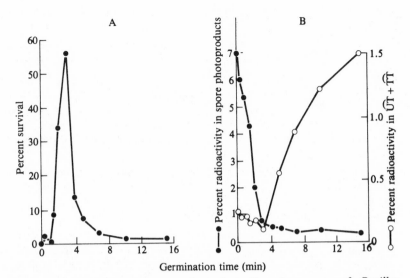

Figure 6.4. Survival and photoproduct formation in spores of *Bacillus megaterium*, UV-irradiated at various times after dilution into germination medium. *Panel A* shows the survival, and *panel B* the formation of spore photoproduct (Sp) and of thymine-containing pyrimidine dimers (PD) after exposure to $320 \text{ J} \cdot \text{m}^{-2}$ of 265-nm radiation. (From R. S. Stafford and J. E. Donnellan, Jr., *Proc. Natl. Acad Sci. U.S. 59*, 822, 1968.)

in *germinating Bacillus subtilis* spores there is a different type of dark repair (spore repair) that apparently monomerizes TDHT in situ.

The importance of both spore photoproducts and pyrimidine dimers for lethality is best demonstrated by the UV resistance pattern displayed during germination (Figure 6.3). The high resistance of the germinating spores is characteristic of only a brief period of transition from the dormant spore to the vegetative cell stage, as illustrated in Figure 6.4. At this stage neither one of the two types of photoproducts is formed to an appreciable extent.

6.3 Biological reasons for lethality

The genetic material of a cellular organism or a virus has to carry out two most essential functions in order to reproduce and perpetuate the biological entity: (1) it must *replicate* itself so that the progeny carry the same genetic specificity possessed by the parent(s), and (2) it must *make available its genetic information* to certain cellular components to provide for all vital phenotypic functions. Consequently, UV photoproducts in DNA could cause lethality by interference with either replication, or information transfer, or both.

There is ample evidence from experimental results obtained in vitro and in vivo that UV-induced cyclobutyl pyrimidine dimers interfere with the DNA replication process. DNA polymerase is apparently unable to deal with a dimer and to recognize it as the two original pyrimidines from which it was formed, and thus terminates synthesis of the daughter strand. There is further evidence that such pyrimidine dimers block the transcription of the genetic material by DNA-dependent RNA polymerase. The result is incomplete messenger RNA molecules, which the ribosomes translate into truncated polypeptide chains. As expected, the blockage at both the replicational and transcriptional level increases with the number of photoproducts, that is, with UV fluence.

It remains an open question whether under certain conditions photoproducts can cause lethality by interfering with replication only (and not with transcription) or with transcription only. Such photoproducts would constitute what we call *solely replicational damage* or *solely functional damage*, respectively. The latter type was invoked on theoretical grounds long ago in connection with results obtained with UV-irradiated phages. To discuss the possibility of its existence, we must realize that only one DNA strand carries the information for phenotypic function, whereas either strand can serve as template in the replication process. Consequently, after low-fluence irradiation it may happen that a critical phenotypic function cannot be carried out, whereas one DNA strand is completely free of dimers and can be replicated.

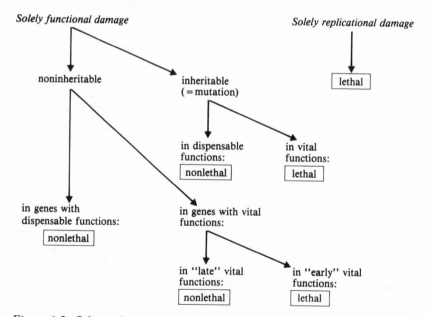

Figure 6.5. Schematic representation of lethal and nonlethal consequences of solely functional, and solely replicational, UV damage (see text.)

In such a case dimers in the functional strand would fit the operational definition of solely functional damage.

Possible biological consequences of solely functional or solely replicational damage are diagrammatically presented in Figure 6.5. Replicational damage must in any case be lethal, irrespective of whether or not it also prevents transcription. In contrast, consequences of solely functional damage (if it existed) would be highly conditional. If functional damage remains unaltered (and thus noninheritable), lethality would depend on whether the affected function were essential or nonessential. But even in a gene with essential function such damage need not be lethal, namely, if the gene product is not required prior to replication, or if its amount in the cell is still sufficient to last until after replication. On the other hand, if as a result of repair functional damage becomes a stable inheritable alteration (*mutation*), it would be lethal if it prevented a *vital* gene function, but not if it prevented a *dispensable* gene function.

Such considerations apply particularly to extracellularly UV-irradiated viruses because, of necessity, any viral gene products are made in the infected cell *after* irradiation. Damage affecting an early function (required *prior* to nucleic acid replication) is likely to cause lethality, in contrast to damage concerning a late function.

6.4 Expression of lethality

Lethality of a UV-irradiated cell or virus is defined as the loss of the ability to reproduce itself. With microbial assay systems, lethality (or inactivation) is recognized by failure to form a macroscopically visible colony or, in the case of viruses, to produce a plaque. Cultured cells of higher organisms usually require microscopic examination for such determination. But all the commonly applied criteria make no distinction between varying morphological or physiological characteristics of inactivated cells or viruses. Irradiated viruses, except after very high UV fluence, usually retain their ability to infect a host cell, but their intracellular development can be arrested at various stages (owing to functional and/or replicational damage) before completion of DNA replication. Irradiated bacteria can show different kinds of behavior related to lethality, which can be recognized microscopically; for example:

(a) After extensive UV exposure an irradiated cell does not show any cellular growth or division.
(b) A cell continues to grow (i.e., increases its mass) but fails to form cross-walls and to divide. The resulting formation of filaments or other giant cell forms is typical of some bacterial strains after low-fluence UV irradiation or various other kinds of injury (see Section 10.3).
(c) UV-irradiated cells continue to grow and undergo a limited number of cell divisions before these activities cease.
(d) Bacterial cells lyse soon after irradiation as a result of inducible lysogeny

or colicinogeny. (Strains not known to be inducible may still carry a defective prophage, whose existence is often difficult to establish.)

On the other hand, lethal UV effects can be demonstrated in certain cases where, by the above definition, a cell survives. An excellent example is the phenomenon of *lethal sectoring*. Its demonstration and investigation require micromanipulation techniques, by which cells are placed individually on a thin agar layer, and the two daughter cells emerging from each of the following four to five divisions are carefully separated. Four classes of cells have been distinguished by this method in UV-irradiated *E. coli* and yeast cells: (1) cells not dividing at all, (2) cells undergoing residual divisions without forming a colony, (3) cells forming a colony but giving rise to inactive de-

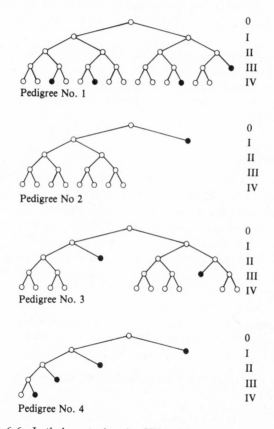

Figure 6.6. Lethal sectoring in UV-irradiated *E. coli* B_{s-1} cells. Typical pedigrees of four surviving cells are shown, which display branches of non-colony-forming cells (●). The Roman numerals represent the number of cell generations. (From K. Haefner and U. Striebeck, *Mutation Res. 4*, 399, 1967.)

scendents (lethal sectoring), and (4) cells forming a colony without lethal sectoring.

The usual plating techniques score the sum of classes (1) and (2) as *inactivated* cells, and the sum of classes (3) and (4) as *survivors*. The existence of class (3) cells demonstrates that survivors are sometimes not free of lethal UV lesions when they divide; examples of pedigrees from such cells are seen in Figure 6.6. A dead-end branch of the pedigree is said to be a *lethal sector* within the progeny. This phenomenon is occasionally observed even with unirradiated cells, but its frequency is much increased after UV- (or X-ray) exposure. Lethal sectors occur most often at the first postirradiation division and then diminish, reaching the spontaneous level after the fourth or fifth cell division.

The basis for lethal sectoring remains to be clarified, but one can reasonably suggest it is a corollary of dark repair processes. In particular, post replication repair (see Section 7.6), which bypasses lethality of unexcised pyrimidine dimers in DNA by recombinationlike processes, is expected to leave behind some daughter cells with unrepaired lethal damage.

7 Recovery and repair

7.1 General considerations

The capability of recovering from injury is common to all organisms. Certainly this is the result of evolution, as possible ways of enhancing an individual's chance for survival are favored by *natural selection*. The means for achieving recovery can vary greatly, depending on the level of biological organization and the kind of injury. Macroscopic regenerative processes, concerning tissues rather than individual cells, have been known for a long time, but evidence for recovery at the cellular level and for the removal of damage from individual biomolecules has accumulated only within the past two or three decades.

The cell component most critical for suffering injury is the genetic material, DNA, not only because of its uniqueness in specificity, but also because of the structural characteristics and giant size of the molecule. It is thus hardly surprising that all known types of molecular repair processes act upon nucleic acids, in particular DNA. Repair processes in other macromolecules have never been discovered, and their existence is, in fact, unlikely for two reasons: First, important molecules not serving as genetic material are usually present in many copies within a cell, making even a considerable reduction in their number tolerable. Second, information for the production of such other molecules is encoded in the genetic material; consequently, replacement of those damaged may be easier for the cell than repair. In the event that replacement is impossible because of damage in the corresponding genome region, the cell would most likely be inviable anyway.

DNA repair mechanisms are now considered fundamental for all organisms; their discovery, investigation, and characterization in molecular terms has been one of the important achievements of ultraviolet radiation biology. Only more recently has it become evident that the same repair processes apply as well to damage by ionizing radiations, photodynamic action, and certain inorganic and organic substances, many of them carcinogenic. Moreover, cells can benefit from the corrective functions of their repair systems in the case of spontaneous DNA damage resulting from thermodynamic causes or inaccuracies in replication, recombination, homologous pairing, and so on. Nevertheless, the extent of repair is most impressive in the case of lethal UV photoproducts in DNA, and some types of repair seem to be specialized for this damage only. This is not surprising in view of the fact that for many organisms ultraviolet light is an ever present environmental factor, and counteraction to its potentially destructive effect must provide a considerable selective advantage.

Terminology

The terms repair, recovery, reactivation, and restoration have been used in publications interchangeably and sometimes with ambiguous meanings. Such usage can hardly be avoided in a rapidly expanding field when there is insufficient information about the basic mechanisms. Although much remains to be clarified, the present state of knowledge permits a less ambiguous use of the terms, which, in accordance with trends in the recent literature, will be applied in the following text.

By *reactivation* (or the synonymous term *recovery*) we mean a *macroscopically observable phenomenon:* in particular the regaining of, by a damaged cell or virus, its capability to propagate and to form a colony, as a result of processes in the absence of which it would fail to do so. The term *restoration* has occasionally been applied in the same sense, but is no longer in general use. In contrast, we define *repair* as the *molecular processes* that are basic for such reactivation. Thus, a recovery phenomenon may have been extensively investigated and well described, but the underlying repair is poorly understood. Also, operational definition of a particular recovery phenomenon could lump together effects resulting from different types of repair. An example is *photoreactivation* (see Section 7.2), which usually is the result of photoenzymatic repair, but occasionally represents light-stimulation of a different type of repair. Conversely, a well-defined repair mechanism might be the basis for a variety of recovery effects observable under different physiological or environmental conditions.

As several fundamental repair mechanisms are now recognized, it is mandatory to specify the type of repair that is meant, provided it is known. The general term dark repair is useful only if it means either the sum of repair mechanisms with the exclusion of photorepair, or the lack of detailed information regarding the mechanism except for the dispensability of light. The term should not be used to mean specifically excision-resynthesis repair, which is the best-characterized type of dark repair.

Detection of recovery effects

Early evidence for a particular recovery process (now called liquid-holding recovery; see Section 8.1) was obtained by Hollaender and Curtis in 1935 for radiation-damaged cells of *E. coli*. In the late 1940s another recovery phenomenon, termed *photoreactivation* (Section 7.2) was described and investigated by A. Kelner for bacteria, and independently by R. Dulbecco for phages. This effect was more suitable for experimental analysis than liquid-holding recovery, and, a decade later, the establishment of its enzymatic nature by C. S. Rupert (1962) had set the stage for an extensive exploration of other recovery phenomena.

This temporal sequence is certainly not fortuitous. Photoreactivation is easier to investigate than any other recovery phenomenon, as it requires an external factor (light), over which the experimentalist has full control. In contrast, other recovery effects depend exclusively on cellular factors; additional experimental parameters may at best enhance or reduce their extent. Thus their demonstration is more difficult, unless there is some a priori expectation for the UV response in the absence of repair. Typically an indication for the existence of such recovery is obtained when either (1) the same irradiated individuals held under different postirradiation conditions give distinctly different survival responses, or (2) the effects in closely related organisms, studied under identical conditions, differ to a considerable extent. The latter condition has been the basis for the isolation of repair-deficient strains, which has contributed greatly to the understanding of repair and recovery phenomena.

Types of repair mechanisms

Apparently the types of mechanisms developed by nature for the repair of DNA are based upon the same principles that govern our repair of technical equipment.

(1) **Reversal of the UV-induced alteration.** This principle is applied in *photoenzymatic repair* (Section 7.2), where UV-induced cyclobutane pyrimidine dimers are monomerized in situ by an enzyme in the presence of near-UV or short-wavelength visible light. An analogous example from the technical world is the repair of an interrupted electrical circuit by reconnecting the separated contacts in situ.

(2) **Replacement of UV-damaged nucleotides.** In the case of *excision-resynthesis repair* (Section 7.4) a UV photoproduct, together with a sequence of adjacent nucleotides, is eliminated from the DNA molecule, and the correct sequence is resynthesized. An analogy from everyday life is the repair of an engine or an electronic gadget by replacing the faulty part.

(3) **Combination of undamaged regions in replicating DNA molecules.** In the cases of *postreplication repair* (or *recombination repair;* Section 7.6) and *multiplicity reactivation* of phage (Section 7.7), undamaged DNA regions are replicated and combined in such a manner than an intact, double-stranded DNA molecule, identical with the original, is formed. This is possible because either strand of the double-stranded DNA molecule carries the information

for replication; and in the case of multiplicity reactivation, even two or more homologous DNA molecules are present within the same cell. The technical analogy is the rebuilding of a functioning engine from two identical, but defective engines, which only requires that no essential part be faulty in both of them.

All DNA repair mechanisms known so far, and probably those still to be discovered, can be accommodated in one of these three categories. However, we should be aware that a recovery effect, as operationally defined, need not *necessarily* be the result of DNA repair. If components other than the genetic material account for part of the UV sensitivity of an organism, the extent of inactivation might likewise be modified by experimental parameters, as the damage could be *bypassed* under one set of experimental conditions but not under another. Although no clear-cut examples for such kinds of recovery can presently be given, they might nevertheless exist.

7.2 Photoreactivation and photoenzymatic repair

Photoreactivation has been broadly defined by Jagger as a "reduction in response to far UV irradiation of a biological system, resulting from a concomitant or post-treatment with non-ionizing radiation." After the explicit description and preliminary characterization of this phenomenon in 1949 by Kelner and by Dulbecco, photoreactivation was observed in other prokaryotes (blue-green algae), as well as in all major categories of eukaryotic organisms (e.g., green algae, fungi, protozoa, echinoderms, arthropods, all major groups of vertebrates, and spermatophytic plants). However, not all species within these taxonomic groups are photoreactivable. For example, among bacteria, *Haemophilus influenzae, Micrococcus radiodurans,* and several members of the genus *Bacillus* lack photoreactivation, and so does the yeast, *Schizosaccharomyces pombe,* whereas baker's yeast (*Saccharomyces cerevisiae*) is highly photoreactivable.

Several kinds of effects are now known to fit the above operational definition of photoreactivation; mostly they are the result of *photoenzymatic repair* (or, for short, photorepair). Others have turned out to be the result of light-stimulated dark repair (*indirect photoreactivation;* see Section 8.4) or of *photochemical monomerization of pyrimidine dimers* (Section 6.2.1), which according to the proposed definition would likewise qualify as a photoreactivation effect.

Photorepair involves a single enzyme (*photoreactivating enzyme* or *photolyase*), which usually requires light energy in the near-UV or violet-blue spectral range (from approximately 310 to 480 nm; *photoreactivating light*) for its effectiveness. However, action spectra differ for different organisms. A first indication for the involvement of a cellular factor, besides light, was the observation that photoreactivation of phage requires illumination of the phage-infected host cell; neither light exposure of the UV-irradiated extra-cellular phage nor separate exposure of cells *and* phage causes recovery. The temperature dependence of photoreactivation suggested the cellular factor to be an enzyme.

Solid evidence for the latter was provided by in vitro studies of Rupert and colleagues in 1958 and thereafter, using UV-irradiated transforming DNA of *Haemophilus influenzae* and extracts from *E. coli* cells. The UV inactivation of transforming activity (Section 5.1) was greatly diminished by exposing such mixtures to photoreactivating light, whereas an identical exposure of irradiated DNA alone, or extracts alone, or both separately, was ineffective. Significantly, cell extracts from *Haemophilus influenzae* (which itself is a nonphotoreactivable bacterial species) failed to promote photoreactivation in vitro. Systematic quantitative study of photoreactivation with an improved assay method (using extracts from the yeast *Saccharomyces cerevisiae* rather than *E. coli*) under varied experimental conditions revealed the following reaction scheme:

$$E + S \underset{k_2}{\overset{k_1}{\rightleftharpoons}} ES \xrightarrow[k_3]{\text{light}} E + P \tag{7.1}$$

The photoreactivating enzyme E combines with its substrate S (a photo-repairable UV lesion) to form an enzyme-substrate complex ES; absorption of light energy photolyzes the complex, thereby releasing the enzyme and con-verting the substrate into the repaired product P. Today it is known that photorepairable lesions are cyclobutyl pyrimidine dimers in DNA, whose monomerization by light in the presence of photoreactivating enzyme has been directly demonstrated.

This in vitro assay system serves several useful experimental purposes. First, it permits the testing of cell extracts, body fluids, and so on, *from any source* for the presence of photoreactivating enzyme activity. As expected, extracts from nonphotoreactivable bacterial species, or nonphotoreactivable (phr^-) mutants of photoreactivable species lack the enzyme activity. Without this test method, it would be much more difficult to establish the presence of photoreactivating enzyme in complex organisms like vertebrates or higher plants, by testing directly their photoreactivability of UV effects.

The *Haemophilus* DNA system can likewise be used to test *any* DNA for the presence of photorepairable damage on the basis of *competitive inhibi-tion*. The repair kinetics observed upon illumination of a mixture of photo-reactivating enzyme and UV-irradiated transforming DNA, depends on the concentrations of the two reactants. The presence, within the same reaction volume, of additional UV-irradiated DNA from another source (nontrans-forming DNA) makes the repair kinetics for the transforming DNA slower because part of the enzyme activity is wasted by interaction with the competing DNA. The extent of slowdown reflects the amount of competing substrate; in the simplest case the repair rate decreases in proportion to the factor $x/(x+y)$, where x and y are the concentrations of the transforming and the competing substrates, respectively.

Thus any treated or untreated DNA or other molecules can be tested for serving as substrate for a photoreactivating enzyme. So far, the only known substrate is UV-irradiated DNA, either double- or single-stranded. Neither *unirradiated* double-stranded DNA nor pyrimidine dimers present in irradiated RNA or in short oligo-deoxyribonucleotides (sequences of less than 9-10 nucleotides) are substrate for the yeast photoreactivating enzyme. However, higher plants possess a photoreactivating enzyme acting on UV-irradiated RNA (see Section 7.2.3).

7.2.1 Extent of photoreactivation

The effectiveness of photoenzymatic repair in a particular biological system can often be expressed by a single numerical index, called the fluence-reduction (or fluence-modification) factor (see Section 4.5). It is a factor (<1.0) by which the effectiveness of the actually applied UV fluence is reduced as a result of maximum photorepair. Consequently, the complementary value (or 1 minus the fluence-reduction factor) expresses the fraction of UV fluence rendered ineffective, which is also called the *photoreactivable sector*.

The photoreactivable sector varies considerably among different biological systems. It can be as high as 0.9 (as in the in vitro *Haemophilus* DNA system), or as low as 0.25 (in the single-stranded DNA phage ΦX 174), but in no case are virtually all lethal UV lesions photorepaired. Therefore, we can formally distinguish between photorepairable and nonphotorepairable UV lesions. Nonphotorepairable lesions could either be photoproducts other than pyrimidine dimers, or pyrimidine dimers that, as a result of secondary alterations or unsuitable adjacent base sequences, are not serving as substrate for the photoreactivating enzyme. Photochemical evidence strongly favors the former suggestion, as virtually all pyrimidine dimers can be monomerized under optimal photorepair conditions.

This conclusion may seem surprising in view of the fact that in DNA other lethal UV photoproducts occur much less frequently than pyrimidine dimers (Section 3.3). We must realize, however, that a very large fraction of pyrimidine dimers is eliminated not only by photorepair, but by various kinds of dark repair processes as well. Whenever photorepair is studied in a biological system that is also subject to dark repair, *we inevitably evaluate the photorepair of only those UV lesions that are not dark-repaired*, even though photorepair usually precedes the other repair. Therefore, if a minority of UV photoproducts are neither photo- nor dark-repairable, whereas most pyrimidine dimers can be repaired either way, the photoreactivable sector determined experimentally must be necessarily lower than the percentage of UV lesions actually photorepaired.

Experimental results confirm this theoretical prediction. In general, one obtains greater photoreactivable sectors (0.80-0.85) in dark-repair deficient *E. coli* cells than in wildtype cells (0.50-0.70). Phage T4, possessing its own excision repair system (Section 7.5), displays a photoreactivable sector of only 0.3-0.35, whereas the mutant T4v^- and the related phage T2, both lacking this repair system, show photoreactivable sectors of 0.65.

The DNA of T-even phages is unusual in containing glucosylated 5-hydroxymethyl cytosine instead of cytosine. Phage mutants lacking glucosyl trans-

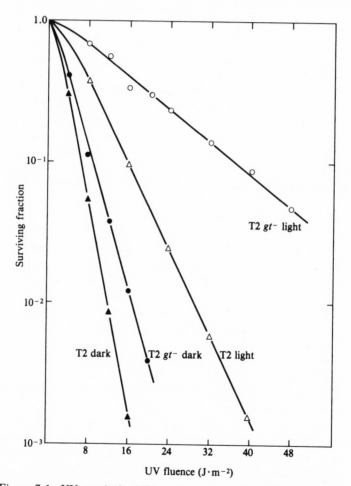

Figure 7.1. UV survival of T2 and nonglucosylated T2 in *Shigella* host cells and their photoreactivation. (From W. Harm, unpublished data.)

ferase (gt^-) cannot glucosylate their DNA, which as a consequence is degraded in their usual host cells. However, such phages can be propagated in certain *Shigella* strains. As Figure 7.1 shows, lack of glucosylation of T2 DNA not only decreases the UV sensitivity of the phage, but also increases considerably the photoreactivable sector.

Phages containing single-stranded DNA are less extensively photoreactivable than those containing double-stranded DNA, presumably because a considerable fraction of lethal photoproducts in single-stranded DNA is not composed of pyrimidine dimers. The photoreactivable sector in these phages hardly exceeds 0.25. Because all of the known bacterial photoreactivating enzymes are DNA-specific, RNA phages are not photoreactivated at all in their host cells. However, RNA-containing plant viruses can repair 20 to 70 percent of their lethal lesions by an RNA-specific photoreactivating enzyme present in higher plants.

7.2.2 The photoenzymatic repair process

As expressed by the reaction scheme (equation 7.1), the first step in photoenzymatic repair is the association of the enzyme with its substrate, a reaction not requiring light. Although this is a reversible step, the dark dissociation rate constant (k_2) is low enough that in cells – as well as under suitable conditions in vitro – the formation of enzyme-substrate complexes is greatly favored. The stability of complexes can be demonstrated by ultracentrifugation of DNA with a limiting amount of enzyme: The enzyme sediments with the DNA if the latter was UV-irradiated, but not if the DNA was unirradiated or was UV-irradiated and previously photorepaired.

The repair step in the strict sense, that is, monomerization of the pyrimidine dimers, is absolutely dependent on light energy. This unusual feature of an enzymatic process permits detailed characterization of the reaction by varying the amount, intensity, and spectral composition of the photoreactivating light. Typically, the light is applied for periods of minutes to an hour, using either monochromatic illumination at wavelengths in the near UV or short visible spectral range, or broad band illumination from daylight or blacklight fluorescent lamps. Although this has led to many important insights into this repair process, the turnover of enzyme molecules during the period of illumination has made a detailed quantitative analysis of the component steps difficult. Only the application of photoreactivating light in the form of an intense flash overcomes this difficulty and permits separate investigation of the left side of the reaction scheme (*complex formation*) and of the right side (*photolysis of the complex*). The following will show how one can determine, from the biological effect, the number of complexes present, as well as their rates of formation and of repair (for review, see Harm et al., 1971).

A simultaneous discharge of several electronic flash units (designed for photographic use) provides sufficient light to photolyze virtually all enzyme-substrate complexes present in a sample at a few centimeters' distance. Therefore, one obtains maximal photorepair if all repairable lesions are complexed during the millisecond flash. If one finds that only a fraction of the repairable lesions are repaired by the flash, one can conclude that only the corresponding fraction of lesions was complexed at that time. Quantitative determination of this fraction from the biological result is based on the following considerations.

We express the repair resulting from a single flash in terms of the *fluence decrement,* ΔF. As indicated in Figure 7.2, $\Delta F = F - F'$, where F is the fluence actually applied, and F' is the (smaller) UV fluence that, without photorepair, gives the same survival as F with photorepair. In other words: ΔF expresses the energy fluence whose effect is annulled by the photorepair treatment, which can be equated with the number of pyrimidine dimers

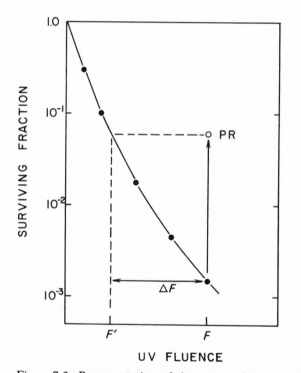

Figure 7.2. Representation of the extent of photoreactivation by the fluence decrement ΔF (see text for details). The open circle, marked PR, represents the survival obtained after exposure to fluence F with subsequent photoreactivation. (From W. Harm, in: *Molecular Mechanisms for Repair of DNA*, Part A, P. C. Hanawalt and R. B. Setlow, eds., Plenum Publ. Co., New York, 1975, pp. 89–101.)

repaired. The steps of evaluation are these: (1) determination of the survival increase as the result of the photoflash; (2) calculation of ΔF from the survival increase; (3) conversion of ΔF into the average number of pyrimidine dimers repaired per individual; (4) equating the number of pyrimidine dimers repaired with the number of enzyme-substrate complexes present at the time of flash. For example, in *E. coli* a fluence of $1 \text{ J} \cdot \text{m}^{-2}$ of 254 nm radiation causes formation of approximately 65 pyrimidine dimers in the chromosome; thus a ΔF of $1 \text{ J} \cdot \text{m}^{-2}$ corresponds to repair of the same number of dimers. Knowing the number of complexes present at a given moment permits *under certain conditions* determination of the number of intact photoreactivating enzyme molecules in the cell or reaction mixture, from which in turn one can establish the reaction rate constants characteristic of the process.

Number of enzyme molecules. In the presence of excess substrate virtually all photoreactivating enzyme (PRE) molecules can be in the form of enzyme-substrate complexes. This condition is met if at further increase in substrate concentration (e.g., by raising the UV fluence at which transforming DNA or cells were irradiated) the number of complexes formed remains constant. This number of complexes equals the number of PRE molecules originally present in the reaction volume.

By this method, the number of photoreactivating enzyme molecules in cells of *E. coli* B derivatives was determined to be of the order of 20. Obviously, mutants containing more enzyme would be recognized by an increased ΔF, that is, by greater survival after a single flash. Figure 7.3 shows measurements of the maximal ΔF that can be obtained with several mutants that were isolated by this criterion. Photoreactivating enzyme concentrations in vitro can be determined in an analogous manner, provided the substrate concentration is known.

Reaction rate constant k_1. The number of complexes determined as a function of time after mixing enzyme and substrate in vitro, or after production of substrate in cell by UV irradiation, reflects the kinetics of complex formation in the dark. It follows from the reaction scheme (equation 7.1) that

$$(d[ES]/dt)_{\text{dark}} = k_1 [E] [S] - k_2 [ES] \tag{7.2}$$

where $[E]$, $[S]$, and $[ES]$ are the concentrations of uncomplexed enzyme, uncomplexed substrate, and ES-complexes, respectively. When flashes are applied in rapid sequence (or continuous illumination is given at *very* high intensity), any complex will be repaired immediately after its formation. Therefore, the term $k_2 [ES]$ becomes negligible (because complexes are so quickly photolyzed) and $[E]$ approximates $[E]_0$, the concentration of free

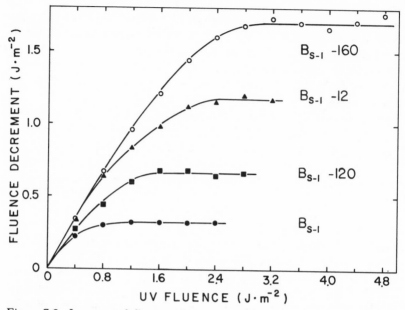

Figure 7.3. Increase of fluence decrement (ΔF) by a single light flash with UV fluence for *E. coli* B_{s-1} and various mutant derivatives. At sufficiently high fluence, ΔF is limited by the average number of photoreactivating enzyme molecules per cell, which is 20, 45, 75, and 110 for the different strains (from bottom to top). (From W. Harm, *Mutation Res. 8*, 411, 1969.)

PRE in the absence of substrate. Accordingly, equation (7.2) can be written:

$$(d[ES]/dt)_{dark} \approx k_1 [E]_0 [S] \tag{7.2a}$$

or in integrated form:

$$\ln([S]_t/[S]_0) \approx -k_1 [E]_0 t \tag{7.2b}$$

Thus application of rapidly sequenced flashes permits determination of the reaction rate constant k_1 from equation (7.2b). The number of lesions before photorepair, $[S]_0$, is known from the fluence applied; the number of lesions remaining after time t, $[S]_t$, equals $[S]_0$ minus the number of repaired lesions (which can be determined from the survival); and $[E]_0$ can be determined as mentioned previously. At room temperature, k_1 is about 10^6 liter mole^{-1} sec^{-1} for the *E. coli* cell and 2.5×10^7 for the in vitro system using *Haemophilus* transforming DNA and yeast photoreactivating enzyme. The large difference between the two values results from the very high viscosity inside the cell, compared to the aqueous in vitro assay mixture. A viscosity increase of the latter by addition of sucrose or glycerol leads to a proportionate decrease of k_1.

Reaction rate constant k_2. When complex formation in the dark has reached equilibrium, the dark dissociation rate constant k_2 for the complexes can be determined by addition of competing substrate at great excess. Under this condition any enzyme molecule dissociating from an established *ES* complex is likely to bind to the competing substrate, and the corresponding loss of original complexes leads to decreasing ΔF obtained by a single flash in the biological assay. At room temperature k_2 is between 10^{-2} and 10^{-3} sec^{-1}, substantiating fully the previously indicated stability of the enzyme-substrate complex.

Photolytic reaction rate constant. Because the photolytic step in the repair reaction depends completely on light energy, the rate constant k_3 (with the dimensions of reciprocal time) is meaningful only when the *fluence rate* L/t and the spectral characteristics of photoreactivating light are specified. If absorption of a single photon by the complex can lead to its photolysis, k_3 can be expressed by $k_p L/t$, where k_p (the *photolysis constant*, with dimensions of a reciprocal fluence) depends solely on the wavelength (or spectral composition) of the photoreactivating light. If virtually all substrate is complexed with enzyme in the dark (a condition that can be tested by the flash method), the fraction of unrepaired complexes $[ES]_L/[ES]_0$ decreases as an exponential function of the light fluence, L:

$$[ES]_L/[ES]_0 = e^{-k_p L} \tag{7.3}$$

The constant k_p, which expresses the relative efficiency of any wavelength with regard to photorepair, is related to the molar absorption coefficient (ϵ) of the complex and the quantum yield Φ of the photolytic reaction. With the following conversion formula the product $\epsilon\Phi$ can be calculated from k_p and λ, using the units stated in brackets:

$$\epsilon\Phi \, [\text{liter} \cdot \text{mole}^{-1} \cdot \text{cm}^{-1}] = k_p \, [\text{m}^2 \cdot \text{J}^{-1}] \cdot \frac{5.2 \times 10^8}{\lambda \, [\text{nm}]} \tag{7.4}$$

A plot of $\epsilon\Phi$ versus the wavelength constitutes an *absolute* action spectrum for the photolytic reaction in photoenzymatic repair, as shown in Figure 7.4. At the maximally effective wavelength region from 366 to 385 nm, the $\epsilon\Phi$ values exceed 10^4 liter $\cdot$ mole^{-1} $\cdot$ cm^{-1} for both yeast and *E. coli* enzyme. Although presently ϵ and Φ cannot be determined separately, the high value of their product tells us that the photolytic process uses the incident light energy very efficiently. Because the quantum yield cannot exceed 1.0, ϵ must be $>10^4$, which means that the absorption of *ES* complexes in the 365-385 nm region exceeds that of pyrimidines at their absorption maxima at 260-270 nm (see Figure 3.1). Because an $\epsilon > 10^5$ is unlikely for organic structures, the quantum yield for photolysis is probably between 10^{-1} and 1.0. A sum-

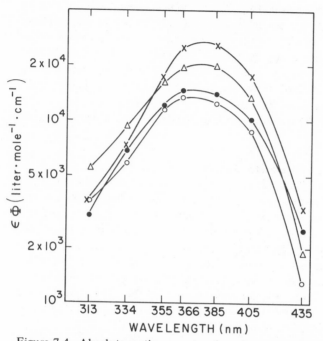

Figure 7.4. Absolute action spectra for photoenzymatic repair of various biological systems. Stationary phase *E. coli* B_{s-1}-160 cells (×); stationary phase cells of *Saccharomyces cerevisiae uvr*l-2, which were mutated to increase the PRE content (•); *Haemophilus* transforming DNA in vitro with yeast PRE either at low purification (○) or at high purification (△). (From W. Harm, in: *Molecular Mechanisms for Repair of DNA*, Part A, P. C. Hanawalt and R. B. Setlow, eds., Plenum Publ. Co., New York, 1975, pp. 89–101.)

mary of relevant findings characterizing the photoenzymatic repair process is presented in Figure 7.5.

7.2.3 Properties of the photoreactivating enzyme

The previously mentioned results and conclusions were obtained with the photoreactivating enzyme present either in its natural cellular environment or in relatively crude protein extracts. Although this work has provided detailed insights into the mechanism of one of the fundamental repair processes, it would certainly be desirable to find out the molecular structure of this unique enzyme in order to understand its biological effectiveness in physical-chemical terms. However, little progress has been made so far in this regard. Repeated attempts to achieve complete purification of the enzyme have met with only partial success. Its low concentration in bacterial and yeast cells (of the order of 10^{-5} to 10^{-6} of total protein) and the tendency

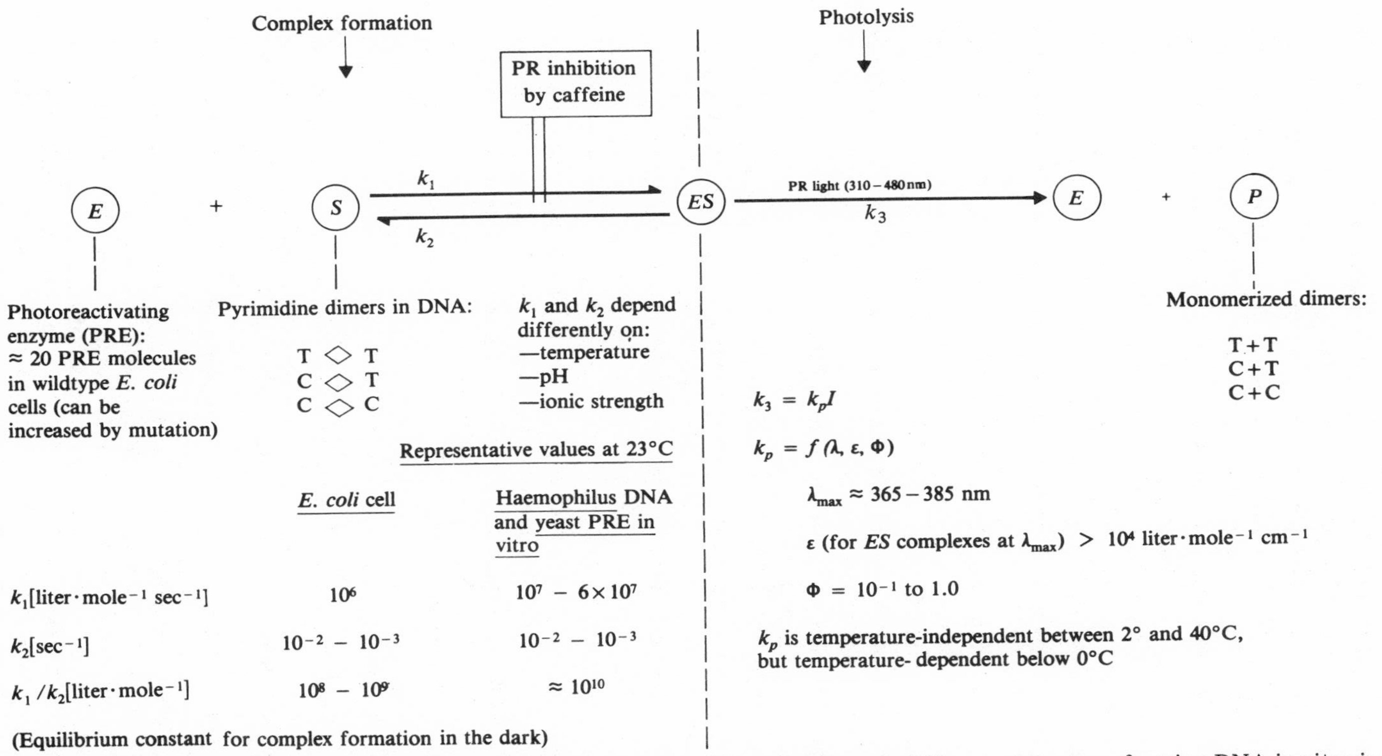

Complex formation

Photolysis

E + S $\rightleftarrows$ (k_1 / k_2) ES $\xrightarrow{\text{PR light (310 – 480 nm)}}_{k_3}$ E + P

PR inhibition by caffeine

Monomerized dimers:

T + T
C + T
C + C

Photoreactivating enzyme (PRE): ≈ 20 PRE molecules in wildtype *E. coli* cells (can be increased by mutation)

Pyrimidine dimers in DNA:

T $\diamondsuit$ T
C $\diamondsuit$ T
C $\diamondsuit$ C

k_1 and k_2 depend differently on:
—temperature
—pH
—ionic strength

$k_3 = k_p I$

$k_p = f(\lambda, \varepsilon, \Phi)$

$\lambda_{max} \approx 365 - 385$ nm

ε (for ES complexes at λ_{max}) > 10^4 liter·mole^{-1}·cm^{-1}

$\Phi = 10^{-1}$ to 1.0

k_p is temperature-independent between 2° and 40°C, but temperature-dependent below 0°C

Representative values at 23°C

	E. coli cell	Haemophilus DNA and yeast PRE in vitro
k_1 [liter·mole^{-1} sec^{-1}]	10^6	$10^7 - 6 \times 10^7$
k_2 [sec^{-1}]	$10^{-2} - 10^{-3}$	$10^{-2} - 10^{-3}$
k_1 / k_2 [liter·mole^{-1}]	$10^8 - 10^9$	$\approx 10^{10}$

(Equilibrium constant for complex formation in the dark)

Figure 7.5. Some characteristics of the photoenzymatic repair of *E. coli* cells and of *Haemophilus* transforming DNA in vitro in the presence of yeast PRE. (From W. Harm, in: *Molecular Mechanisms for Repair of DNA*, Part A, P. C. Hanawalt and R. B. Setlow, eds., Plenum Publ. Co., New York, 1975, pp. 89–101.)

to lose its activity during the biochemical procedures were the major difficulties, but they have been overcome recently.

The yeast enzyme molecule seems to consist of two nonidentical subunits of roughly 55,000 and 80,000 daltons and a low-molecular-weight chromophore, while the *E. coli* enzyme consists of a single 35,000 daltons unit and a cofactor. The very high absorbance of the enzyme-substrate complex in the 360–385 nm spectral region, which is evident from the biological results, apparently does not hold for the free enzyme. This might be an effective mode of protection against photochemical destruction of the enzyme by sunlight. Another characteristic difference between the *complexed* photoreactivating enzyme and the free enzyme is in the sensitivity to heat and heavy metal ions. In the complexed state the inactivation rate by these agents is reduced to about one 1/10.

Biological experiments of the type described in Section 7.2.2 have shown that the complex formation and the stability of the complexes, characterized by both k_1 and k_2, are greatly affected by the *ionic strength, pH,* and *temperature.* Depending on the composition of the buffer, k_1 shows a narrow maximum at ionic strengths between 0.15 and 0.25, whereas k_2 increases with increasing ionic strength from 0.05 to 0.25. Both k_1 and k_2 increase with pH ranging from 6.0 to 7.5, and, as expected, both show positive temperature dependence corresponding to activation energies of approximately 9–11 kcal · mole^{-1} for k_1, and 4.5–5.5 kcal · mole^{-1} for k_2. The photolytic properties depend relatively little on these parameters, except that at temperatures below 0°C the $\epsilon\Phi$ values drop continuously to reach at –196°C about 2.5 percent of their normal level.

All these properties are characteristic for the *Saccharomyces* photoreactivating enzyme. However, photoreactivating enzymes from other organisms may have different properties, as suggested by the different action spectra displayed in Figure 7.6. Such spectra are important for determining safe light

Figure 7.6. Action spectra for photoenzymatic repair with photolyase from various organisms. (The ordinate scale is logarithmically divided. Each curve is normalized so as to coincide with the ordinate value of 1.0 at its peak.) (1) *Staphylococcus epidermidis*: PR of UV-irradiated cells (from M. Ikenaga et al., *Photochem. Photobiol. 11*, 487, 1970). (2) *Neurospora crassa*: Action of PRE-containing extracts on UV-irradiated *Haemophilus* transforming DNA (from C. E. Terry and J. K. Setlow, *Photochem. Photobiol. 6*, 799, 1967). (3) *Streptomyces griseus*: PR of UV-irradiated conidia (from J. Jagger et al., *Photochem. Photobiol. 12*, 185, 1970). (4) *Phaseolus vulgaris* (pinto bean): Action of PRE-containing extracts on UV-irradiated *Haemophilus* transforming DNA (from N. Saito and H. Werbin, *Radiation Botany 9*, 421, 1969). (5) *Saccharomyces cerevisiae*: PR of UV-irradiated cells (from H. Harm, in: *Photochemistry and Photobiology of Nucleic Acids*, S. Y. Wang, ed., Vol. II, Academic Press, New York, 1976, pp. 219–63. (6) *Anacystis nidulans*: Action of PRE-containing extracts on UV-irradiated *Haemophilus* transforming DNA (from N. Saito and H. Werbin, *Biochemistry 9*, 2610, 1970.)

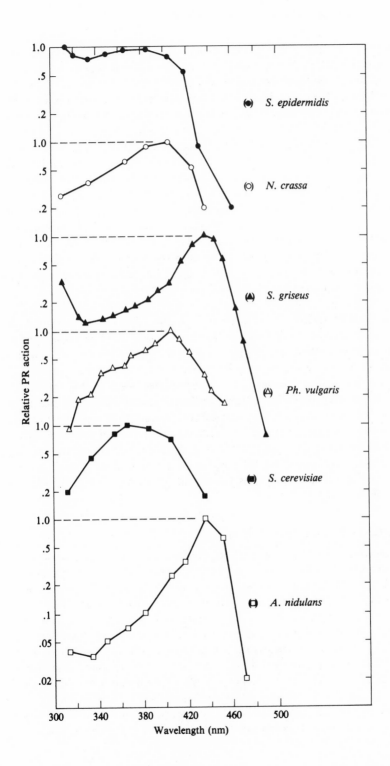

sources in the laboratory for experimental manipulation with samples not to be photorepaired. Working safely with yeast and *E. coli* requires wavelengths >500 nm, a condition satisfied by the use of General Electric Gold fluorescent lamps. Other organisms (e.g., *Streptomyces griseus* with a peak of PR action at 435 nm) may require orange or red fluorescent lamps to prevent photoreactivation.

Absence of photorepair is sometimes desired in experimental work involving near UV or visible light. In certain cases photorepair can be considerably inhibited by 0.1 M caffeine, which competes with the enzyme for binding to its substrate. However, the least objectionable way of avoiding photorepair is the use of nonphotoreactivable mutants (*phr⁻*), which have been isolated from various *E. coli* strains and *Streptomyces griseus*, and can probably be isolated wherever necessary. Evidently the photoreactivating enzyme is not a vital cell component, so that its absence has no recognizable phenotypic effects other than those related to UV repair. This supports the view that the latter is its predominant or sole function.

RNA-specific photoreactivating enzyme. Very little is known about the properties of RNA-photoreactivating enzyme found in higher plants. Like its DNA counterpart, it repairs pyrimidine dimers (mainly of the uracil-uracil or uracil-cytosine type) in RNA, and thus permits photoreactivation of UV-irradiated, RNA-containing plant viruses after infection. For some time, a notable exception in this regard was the rod-shaped *tobacco mosaic virus*. It turned out later, however, that tobacco mosaic virus can be fairly well photoreactivated when it was UV-irradiated in the form of naked viral RNA. The difference is due to the fact that in the intact virus the RNA molecule is held rigidly in an extended position by a helical array of protein units wrapped around it, which does not permit formation of pyrimidine dimers (see Section 5.2).

7.3 Host-cell reactivation of viruses

In the mid-1950s, Garen and Zinder discovered that UV inactivation of the *Salmonella* phage P22 shows a strong dependence on the conditions in their host cells. If the cells had been heavily UV-irradiated or X-rayed prior to infection, phage survival was considerably lower than after infection of unirradiated host bacteria. Figure 7.7 indicates that with increasing fluences at which the cells are preirradiated, a rising percentage of the phage show the high UV sensitivity. Whereas the *intrinsic* UV sensitivity (see Section 4.2.1) of P22 infecting unirradiated host cells is considerably lower than that of phage T2, this difference disappears in heavily irradiated host cells, in which the T2 survival is the same as in unirradiated cells. The low intrinsic UV sensitivity of P22 can be attributed to host-cell reactivation of the phage, which takes place in unirradiated, but not heavily irradiated cells. Host-cell

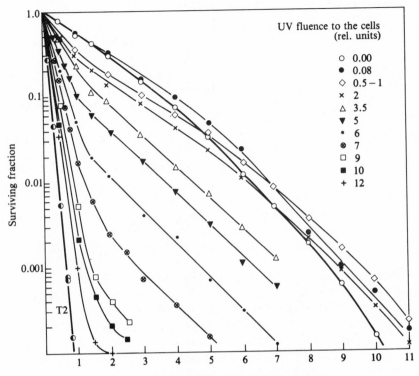

Figure 7.7. UV survival of phage P22, infecting *Salmonella typhimurium*. UV exposure of cells and phages is expressed in relative units of fluence. The survival curve of phage T2 is shown for comparison. (From A. Garen and N. D. Zinder, *Virology 1*, 347, 1955.)

reactivation is ordinarily observed in temperate phages (e.g., P22, lambda, and many others), and moderately virulent phages such as T1, T3, and T7, but is not found with the highly virulent T-even phages and T5. It does not occur in phages containing single-stranded DNA or RNA (see Figure 4.2).

Originally, the mechanism of host-cell reactivation was assumed to be genetic recombination between the phage and a *homologous* region of the host chromosome. This interpretation became questionable several years later, when it was found that the very UV-sensitive mutant strain B_{s-1} of *E. coli* lacks the ability to host-cell-reactivate phage. Thereafter, a variety of other *E. coli* mutants specifically isolated for their failure to reactivate phages were likewise found to be highly UV-sensitive themselves. Evidently, the factor responsible for host-cell reactivation of phage affects in a similar manner UV damage in the bacterial chromosome itself. This is easily understood if host-cell reactivation of phage is the result of cellular enzymatic repair that affects UV lesions in both phage and bacterial DNA, and for which the mutant strains are defective.

Subsequent experimental work confirmed this concept. The essential mechanism underlying host-cell reactivation is bacterial *excision-resynthesis repair,* a complex sequence of enzymatic processes described in Section 7.4. Comparison of phage survival in excision-repair proficient (uvr^+) and deficient host strains shows fluence modification factors of approximately 0.15 to 0.20, indicating that 80 to 85 percent of otherwise lethal phage lesions are repaired. This is a considerable percentage, but still much below that achieved by excision repair in UV-damaged bacterial DNA itself.

Other cellular repair systems like *recombination* or *postreplication* repair (see Section 7.6) play a minor role, if any, in host-cell reactivation. For example, the UV survival of phage lambda is lower in recombination-repair deficient ($recA^-$) cells than in $recA^+$ cells (see Figure 7.20), but this difference is small compared with that between uvr^+ and uvr^- host cells. Other host-cell-reactivable phages, like T1, T3, and T7, are virtually unaffected by the $recA$ function.

Host-cell reactivation is almost completely inhibited in the presence of 2 mg caffeine per ml in the agar medium, which still permits growth of many bacterial indicator strains and plaque formation by both unirradiated and irradiated phage. A similar inhibition is observed with acriflavine at concentrations of 1 to 4 μg/ml. Without inhibition, the extent of excision repair leading to host-cell reactivation apparently varies within the population of phage-infected cells, as indicated by the upwardly concave survival curves found with phages T1, T3, T7, HP1, and others (see Figure 4.10). Such curves are not observed when excision-repair deficient host cells are infected.

Figure 7.8 shows that *partial* repair inhibition accentuates the two-component character of the curves, by lowering the fraction of extensively host-cell-reactivated phage individuals and reducing the extent of repair in the remainder. Only at very high inhibitor concentrations or after extensive preirradiation does the phage survival resemble that in an excision-repair deficient host because the shallow curve component would be below the survival levels usually considered. Analysis of survival curves obtained under conditions of partially inhibited host-cell reactivation suggests that the shallow curve portion represents phage-infected cells in which the repair is immediate and extensive. This requires, in addition to the excision repair enzymes, a cellular factor inhibitable by preirradiation, caffeine, or acriflavine.

It has been an open question for a long time why T-even phages and T5 (the latter containing only the usual four bases in DNA and no glucose) do not take advantage of the excision repair system provided by the host cells. Recent work indicated that these phages produce a protein *interfering* with repair, which – if otherwise vital for these phages – would make their lack of host-cell reactivation inevitable. Failure of host cells to reactivate phages con-

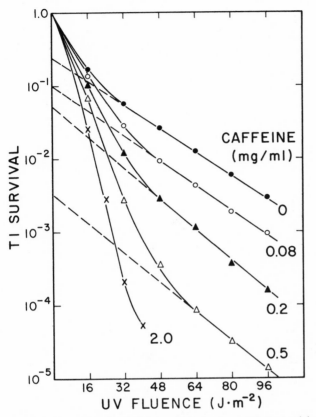

Figure 7.8. Survival of UV-irradiated phage T1 in excision-repair proficient host cells (*E. coli* B/r) in the presence of various concentrations of caffeine. (From W. Harm, *Mutation Res. 25*, 3, 1974.)

taining single-stranded DNA or RNA follows from the mechanism of excision repair, which relies on the double-strandedness of the nucleic acid molecule (Section 7.4).

Recognition of the host-cell reactivation phenomenon in phage and clarification of the underlying repair processes prompted search for an analogous recovery in other viruses, notably animal viruses containing double-stranded DNA, for example, SV40, *herpes simplex,* and the *adenoviruses*. Mammalian cells are capable of excision repair; the significance of this repair is evident from its absence in cases of the hereditary disease xeroderma pigmentosum, where sunlight exposure results in multiple skin carcinomas and often death of the individual. The survival of UV-irradiated viruses, plated on cell strains from such patients, is much reduced compared with plating on excision-repair proficient cells. Examples are shown in Figure 7.9, where the same UV-

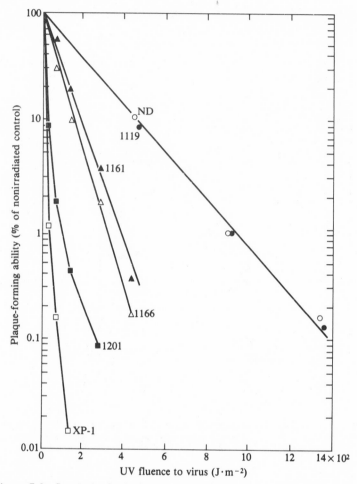

Figure 7.9. Survival of adenovirus 2 infecting human fibroblasts from either "normal" cell lines (ND; 1119) or from various cell lines originating from patients with varying degrees of xeroderma pigmentosum. (From R. Day, in: *Molecular Mechanisms for Repair of DNA, Part B*, P. C. Hanawalt and R. B. Setlow, eds., Plenum Publ. Co., New York, 1975, pp. 747–52.)

irradiated samples of adenovirus 2 are plated with human fibroblasts from a "normal" cell line and from various lines of xeroderma origin. The reduction in the extent of host cell recovery depends on the cell line and is most pronounced in XP-1, where the symptoms of the disease are most severe. In this case, the mean inactivation fluence $(F_{0.37})$ for the virus is $10 \text{ J} \cdot \text{m}^{-2}$, corresponding to the formation of one to two pyrimidine dimers per genome, which suggests that little or no other repair contributes to the host cell reactivation of this virus. Comparison with the survival curve obtained in normal fibroblasts shows that the latter repair at least 95 percent of the lethal UV lesions in the viral DNA.

7.4 Excision-resynthesis repair

The search for UV-sensitive bacterial mutants lacking the capability to host-cell-reactivate phages, and the study of UV effects in these mutants, has led to the discovery of a fundamental dark repair mechanism, which is of great importance not only in prokaryotes but in plants and animals as well. In 1964, R. B. Setlow and W. L. Carrier, and at about the same time R. P. Boyce and P. Howard-Flanders, compared UV-irradiated cells of host-cell-reactivating (*hcr*$^+$) and non-host-cell-reactivating (*hcr*$^-$) *E. coli* strains regarding the fate of pyrimidine dimers in their DNA. They found that, within an hour after irradiation, pyrimidine dimers are rapidly eliminated from the DNA of *hcr*$^+$ cells, but not from the DNA of *hcr*$^-$ cells. Dimer removal is achieved by enzymatic *excision* of oligonucleotides from one DNA strand, whose integrity is then restored by specific *resynthesis* of the excised nucleotide sequence, using the information provided by the complementary DNA strand. This process has, therefore, been termed excision-resynthesis repair, or simply excision repair.

7.4.1 Mechanism of excision repair

Subsequent work in various laboratories has revealed more details of this repair mechanism. A schematic representation is shown in Figure 7.10. The first step is *cleavage* (nicking or incision) of the phosphor-diester backbone adjacent to a dimerized nucleotide pair. This reaction is carried out by a *repair endonuclease* (UV endonuclease or correndonuclease), which recognizes the alteration and cleaves the polynucleotide strand on the 5′ side of the dimer. Although this step leaves the double-helical DNA structure essentially intact, its occurrence is easily demonstrated by *alkaline* sucrose gradient sedimentation, where denaturation of the DNA molecules reveals the shortened single strands. The next step in the repair process, *excision,* is evident from the progressive disappearance of the pyrimidine dimers from the (acid-precipitable) DNA and their reappearance in the form of (acid-soluble) oligonucleotides. It is a logical postulate that the excised nucleotide sequence must be replaced by *resynthesis*, for which only the intact complementary strand can serve as a template. This step, first demonstrated in experiments by Pettijohn and Hanawalt in 1964, soon after the discovery of dimer excision, is called *repair replication* (or repair-synthesis). It is distinct from the regular DNA replication by occurring in the form of many short, single-stranded nucleotide sequences, each filling the gap resulting from the removal

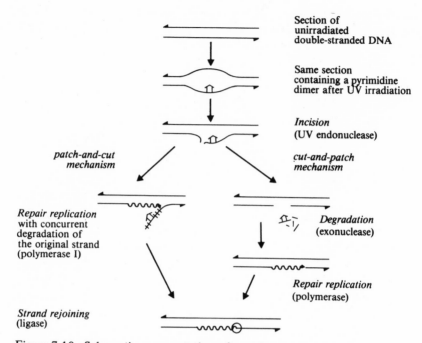

Figure 7.10. Schematic representation of reaction steps involved in excision-resynthesis repair (see text.)

of a dimer and adjacent nucleotides. The final repair step, called *rejoining,* integrates the newly synthesized material into the DNA molecule by covalent binding.

Evidently the mechanism of excision repair is more complex than that of photorepair because it involves at least four enzymatic activities, which must be performed in a coordinate fashion. Because such coordination need not be identical for cells from different organisms, or even for a particular type of cell under different experimental conditions, one must not be surprised that investigation of excision repair processes at the molecular level in different laboratories has sometimes led to discrepant findings. We will not attempt to discuss them here, but rather will restrict our discussion to firmly established facts.

Incision step. The complexity of the excision repair mechanism would make it a reasonable expectation that repair-deficiency mutations can block the sequence of repair steps at different points, much as auxotrophy mutations can block a metabolic pathway at different reactions. However, all excision-repair deficient *E. coli* mutants isolated soon after discovery of this repair turned out to be unable to excise the dimers. The mutations were mapped in three genes called *uvrA, uvrB,* and *uvrC,* located in different chromosomal regions. The *uvrA* and *B* mutants lack UV-endonuclease activity and are thus unable to perform the incision step, whereas *uvrC* mutants apparently possess endonuclease, but in the absence of the *uvrC* gene product excision fails because a nick is immediately ligated.

All three types of mutants are completely unable to host-cell-reactivate phages, and their own UV sensitivity is much higher than that of wildtype cells (Figure 4.7, panels B and C). In order to determine the fraction of bacterial DNA lesions that are excision-repaired in the absence of any other repair, we compare the *uvrA⁺recA⁻* curve in Figure 4.7, panel B, with the *uvrA⁻recA⁻* curve in panel A. For the survival range from 100 to 5 percent, where the two curves have similar shapes, the fluence modification factor is about 0.035, indicating that 96.5 percent of the lesions are excision-repaired. Comparison of wildtype (*uvrA⁺recA⁺*) survival with *uvrA⁻recA⁺* survival (panels C and B of Figure 4.7) shows fluence modification factors from 0.025 to 0.05, suggesting that the overlapping *recA⁺*-controlled repair hardly affects the estimated fraction of excision-repaired lesions.

Excision and resynthesis steps. For the excision and resynthesis steps different exonuclease and polymerase activities present in *E. coli* cells have been considered. Apparently the prime role is played by *DNA polymerase I* (Kornberg polymerase), which also exhibits $5' \rightarrow 3'$ exonucleolytic activity. This enzyme, as it polymerizes, degrades the nicked, dimer-containing DNA-strand, thereby translating the nick in the $5' \rightarrow 3'$ direction. *polA* mutants lacking the polymerizing function of this enzyme are considerably more UV-sensitive than the wildtype strains from which they are derived, but are by far not as sensitive as *uvrA, B,* or *C* strains. Similarly, host-cell reactivation of UV-irradiated phages is reduced in *polA* cells, but not absent. Therefore, we conclude that the extensive excision repair characteristic of wildtype cells requires activity of intact polymerase I, but in its absence some excision repair still occurs. In the latter case, probably another DNA polymerase (II, or III), not normally involved in this type of repair, carries out the function less effectively.

In contrast, *polA* mutants lacking the exonuclease rather than the polymerase function are little affected in repair of their own DNA or that of infecting phages. Thus, either substitution for the lacking exonucleolytic function of polymerase I by a different exonuclease (for example, exonuclease III or VII) is not disadvantageous, or the excision is not even normally carried out by the exonuclease activity of polymerase I.

For visualizing the sequence of events during the excision and resynthesis phase, mainly two possibilities have been considered. They have been referred to as *patch-and-cut* and *cut-and-patch* mechanisms, as indicated in the scheme (Figure 7.10). However, one can easily envisage intermediate situations, which are neither one nor the other. Obviously patch-and-cut is the mode of repair when both excision and repolymerization are carried out by the Kornberg polymerase. The alternative cut-and-patch mechanism, which temporarily leaves a single-strand gap of appreciable length, may nevertheless prevail when DNA polymerase I is nonfunctional. It is not at all unlikely that in cells studied under metabolically different conditions (for example, logarithmic growth versus starvation in buffer) the excision and resynthesis steps involve different enzymes and are differently coordinated.

Rejoining. Covalent binding of the newly synthesized nucleotide sequence with the original DNA strand is carried out by *polynucleotide ligase,* the same enzyme that is involved in replication and recombination of the undamaged genome.

Although the principal repair steps – incision, excision, resynthesis, and ligation – seem to characterize excision repair in general, differences exist in various organisms regarding details of the reactions and the controlling enzymes. For example, resynthesized nucleotide sequences are usually much longer (>100 nucleotides) in mammalian cells than in *E. coli* cells (10–30 nucleotides), unless the latter are deficient in polymerase I. UV-endonuclease of *E. coli* differs from that of phage T4, or *Micrococcus luteus* (where actually two separable enzyme activities of this kind have been demonstrated). Furthermore, the complexity of the excision repair system renders it susceptible to great variations in its effectiveness due to several experimental and natural parameters (see Chapter 8).

As mentioned for host-cell reactivation of bacteriophages, excision repair can be inhibited by caffeine (at concentrations up to 2 mg/ml) or acriflavine (up to 4 µg/ml). Caffeine binds to single-stranded DNA, and to UV-irradiated (but not unirradiated) double-stranded DNA, and has been shown to interfere with photoenzymatic repair at the level of enzyme-substrate complex formation. Thus it is likely that interference with excision repair is due to its binding at locally denatured DNA regions, which are the result of pyrimidine dimer formation. In contrast to phage DNA, bacterial DNA is inhibited by caffeine to greatly varying extents in different strains, and inhibition is usually far from being complete. In wildtype cells, the UV survival is usually much less affected than by *uvr⁻* mutations; the same holds for acriflavine, which can both intercalate between the base pairs of double-stranded DNA, and bind on the outside.

As the DNA of UV-irradiated phages entering a host cell can be excision-repaired, so can be bacterial transforming DNA, or infectious viral DNA. This is best demonstrated by comparative infection of excision-repair proficient and deficient cells with the same irradiated material. Moreover, in repair-proficient *Haemophilus* cells the survival function of UV-irradiated transforming DNA varies greatly for different markers. The differences are much reduced under conditions of repair inhibition, suggesting that marker-characteristic properties determine the extent of excision repair. Because only one DNA strand of *Haemophilus* transforming DNA is integrated into the recipient cell genome, whereas the repair requires a double-stranded DNA structure, the time of integration is likely to be relevant for the extent of repair, and could explain the differences in marker survival.

Excision repair studied in vitro. The requirement of double-strand-edness for excision repair is implicit from its mechanism, and has been directly demonstrated by comparison of the UV survival of single-stranded and double-stranded infectious DNA of phage ΦX 174 (see Figure 5.3). The single-stranded form shows identical curves for infection of either repair-proficient or repair-deficient host strains. In contrast, double-stranded DNA (replicative form) infecting repair-proficient cells is far more resistant than when infecting repair-deficient host cells.

UV-irradiated replicative form of ΦX 174 DNA has been used for demonstration of the incision step in repair in vitro by UV-endonuclease-containing cell extracts from *Micrococcus luteus*. Infection of *E. coli uvr⁻* cells with treated DNA results in much higher phage survival than infection with untreated control DNA. If the endonucleolytically cleaved DNA is treated in vitro with *E. coli* polymerase I, nucleotide sequences are removed and resynthesized, and upon addition of polynucleotide ligase the DNA can be reconverted to the original covalently closed circles.

7.4.2 General significance of excision repair

Although excision repair has been investigated mainly in *E. coli* and some other bacterial species, it appears to be of general importance in the world of living organisms. No evidence has been obtained so far for its absence in any species where it has been searched for. Because excision repair deficiency causes unusual UV sensitivity in bacterial cells, established UV survival curves for strains of many bacterial species suggest that all of them possess an excision repair system. Furthermore, the occurrence of this type of repair has been explicitly shown in mammalian cell cultures, including those of human origin, although the repair kinetics (in terms of *percentage* of dimers removed per unit time) is evidently slower than in bacteria.

Compared with photoenzymatic repair, which deals essentially with pyrimidine dimers, excision repair is less specific with respect to the kinds of DNA alterations concerned. Besides pyrimidine dimers, spore photoproducts can be excised, and in the same manner several types of chemical DNA lesions can be eliminated. Examples are alterations by mono- and bifunctional alkylating agents like methyl-methanesulfonate (MMS), ethyl-methanesulfonate (EMS), di-(2-chloroethyl)sulfide (mustard gas), mitomycin C, and others. Excision repair has also been found effective for lesions caused by ionizing radiation, 4-nitroquinoline-N-oxide (4-NQO), and photodynamic DNA damage caused by visible light in the presence of acridine dyes or psoralen. However, the repair of damage is often less extensive than after UV irradiation. This is not surprising if ultraviolet light as a natural environmental factor has been the major selective force in establishing and maintaining this repair mechanism. Nevertheless, we can consider it a general "DNA-maintenance system" for the cell, whose predominant function is the recognition and correction of various kinds of irregularities, UV damage included. The latter has been most essential not only for the discovery of this repair system, but also for the isolation of a great variety of mutants, which are of basic importance for a detailed investigation of its mechanism.

7.5 Phage-controlled repair: v-Gene reactivation

The UV survival of T-even phages is unaffected by the *uvr* genotype of their host cells. This shows that they are not host-cell-reactivated by bacterial excision repair. Instead, T4 possesses an excision repair system of its own, which involves the product of phage gene v^+. This has the result that phage T4 is considerably more UV-resistant than T2 and T6, although the three types of T-even phages are closely related genetically and serologically, very similar in their DNA content and composition, and morphologically indistinguishable.

About 30 years ago Luria concluded, from hybridization experiments between T2 and T4, that the sensitivity difference must be caused by a single gene. Later on, Streisinger mapped this gene, and found that it likewise controls the extent of photoreactivation, in which the two phage types also differ. Evidence that T4, in contrast to T2, possesses a dark repair mechanism that greatly overlaps with photoenzymatic repair was obtained in 1958 by Harm: In mixed infection of cells with T2 and *heavily* UV-irradiated T4, phage T2 shows the more resistant UV survival curve and the reduced level of photorepair characteristic of T4, as seen in Figure 7.11. The conclusion that T4 contains a repair gene in its active state (v^+), which is absent or defective in T2, was supported by the isolation of T4v^- mutants, whose UV effects resemble those of T2 and T6. The resulting recovery effect was henceforth called v-gene reactivation; a recent review of this phenomenon and the underlying repair was published by Friedberg (1975).

T4-mediated excision repair resembles *E. coli* excision repair, except that it appears to be fairly specific for UV damage. However, more recently Benz and Berger found that the v-gene system is also capable of recognizing and excising single-strand loops in DNA, which can be produced experimentally by annealing two complementary single DNA strands differing by a deleted or duplicated nucleotide sequence (*heteroduplex repair*). Thus the phage-controlled repair system may be generally capable of correcting irregularities in DNA occurring in the process of recombination and/or replication. The

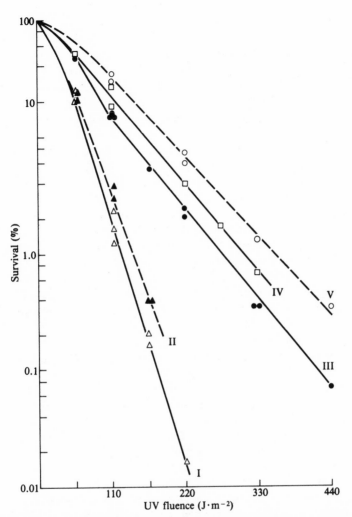

Figure 7.11. UV survival of phages T2 and T4 infecting *E. coli* B cells under the following experimental conditions: (I) single infection of cells by irradiated T2 only; (II) single infection by irradiated T2 of cells simultaneously infected with heavily irradiated (330 J · m^{-2}) T2 at a multiplicity of 3; (III) single infection by irradiated T2 of cells simultaneously infected with heavily irradiated (330 J · m^{-2}) T4 at a multiplicity of 3; (IV) single infection of cells by irradiated T4 only; (V) single infection by irradiated T4 of cells simultaneously infected with heavily irradiated (330 J · m^{-2}) T4 at a multiplicity of 3. (From W. Harm, *J. Cell. Comp. Physiol. 58*, Suppl. *1*, 69–77, 1961.)

perfectly good viability of unirradiated T2 phage and T4v^- mutants shows, however, that such a correcting mechanism cannot be of vital importance for the phage, suggesting that its prime concern is the UV repair function.

The v-gene product is a UV-endonuclease (called endonuclease V) cleaving the DNA strand at the 5' side of a pyrimidine dimer, which was purified and characterized independently in the labs of Friedberg and Sekiguchi. As in *E. coli*, T4 mutants lacking this endonucleolytic activity are completely excision-repair deficient, but it is still an open question whether other enzymes involved in this repair are of bacterial or phage origin. T4 produces a UV-exonuclease, but repair-deficient mutants lacking this activity have not yet been found. On the other hand, irradiated T4 is slightly more sensitive when it infects *polA*$^-$ cells (lacking DNA polymerase I activity) rather than *polA*$^+$ cells, which suggests that repair replication is normally carried out by the bacterial polymerase I rather than by the T4 DNA polymerase.

Excision of pyrimidine dimers from UV-irradiated T4 DNA has been demonstrated in vitro in the presence of magnesium ions and extracts from T4-infected cells. It is not specific for T-even DNA (which contains glucosylated hydroxymethyl cytosine instead of cytosine); dimers in UV-irradiated *E. coli* DNA are likewise excised in vitro by the same repair system. Correspondingly, UV-irradiated *E. coli uvr*$^-$ cells can undergo v-gene reactivation in vivo under appropriate experimental conditions: They show a 10- to 100-fold survival increase after infection with heavily preirradiated (200–500 J·m^{-2} at 254 nm) T4v^+ phage, an effect that is not observed if the infecting phage is T4v^-. The heavy irradiation is required to inactivate in most phage particles their fairly UV-resistant host-killing ability, while still leaving intact the v^+ gene function whose UV sensitivity is only about $\frac{1}{100}$ that of the phage as a plaque former. Similarly, host-cell reactivation of irradiated phage T1 can be simulated in excision-repair deficient *E. coli* cells by the presence of a T4v^+ phage. Again, the latter must be heavily preirradiated, in order to avoid complete exclusion of phage T1 in mixed infection with T4.

v-Gene reactivation of T4 is somewhat less extensive than host-cell reactivation of phages by cellular excision repair. Comparison of T4v^+ and T4v^- survival curves indicates repair of approximately 55 percent of lethal lesions; however, this value is probably too low because of the overlap with another T-even-specific type of recovery: *x-gene reactivation* (Section 7.10). The slopes of T4 survival curves in the absence, of *x*-gene reactivation indicate 60 to 65 percent v-gene repair for T-even DNA, whereas v-gene repair of phage T1 in an excision-repair deficient host strain corresponds to approximately 70 percent repair.

7.6 Recombination (or postreplication) repair

In 1965, soon after the discovery of excision repair, Clark and Margulies isolated two *E. coli* K12 mutants, which produced extremely low numbers of recombinants when male (*Hfr*) chromosome fragments were transferred into female (F⁻) cells by conjugation. A considerable number of such *recombinationless* (or *rec⁻*) strains have subsequently been isolated in various laboratories; the mutations map at various distinct regions of the *E. coli* chromosome (gene *recA, recB*, etc. . . . *recL*). Most such mutants affect also the sensitivity to UV radiation, and of particular interest here are the *recA* mutants. Their UV sensitivity resembles roughly that of *uvrA, B*, and *C* mutants, but they host-cell-reactivate phage and excise pyrimidine dimers.

  recA mutants are defective in a DNA repair mechanism involving genetic recombination or recombination-related processes, which is called recombination repair (Rec repair) or postreplication repair. That this repair mechanism is distinct from excision repair is evident from the fact that double mutants of the type *uvrA⁻ recA⁻* are 30 to 40 times more UV-sensitive than the corresponding *uvrA⁻* strain (Figure 4.7). In fact, such double-mutants are the most sensitive *E. coli* strains ever observed. The mean inactivation fluence at 254 nm is no more than $0.02 \text{ J} \cdot \text{m}^{-2}$, causing formation of approximately one pyrimidine dimer per *E. coli* chromosome. This is consistent with the belief that these cells are unable to carry out *any* dark repair of UV lesions, although the remaining intact functions can still repair some other kinds of DNA damage. The fundamental difference between excision and recombination repair is also emphasized by the slight effect of the latter on UV-irradiated phages. For example, infection of *recA* cells reduces the survival of phage lambda only slightly and that of T1 hardly at all. It is a reasonable assumption that postreplication repair processes resembling Rec repair of *E. coli* are as widespread among organisms as excision repair.

  Mutant strains resembling in their properties *E. coli recA* have been found in a considerable number of bacterial species, in yeast, and apparently in mammalian cell lines.

7.6.1 Mechanism of recombination repair

Fewer details are known about the mechanism of Rec repair than about excision repair, but its principal nature is now well established. Unlike excision repair, at least partial DNA replication is required before recombination repair can take place, which explains the alternate term *postreplication repair*. This term is now frequently used, particularly for organisms less well investigated than *E. coli*, where the requirement of regular DNA replication is sometimes the only criterion applied. However, in *E. coli* the term is less desirable because of its possible confusion with repair replication (the resynthesis step in excision repair), and also because it is a matter of semantics to decide whether the repair operates *post* replication or *during* replication.

  In 1968 Rupp and Howard-Flanders obtained first evidence for the mechanism involved in Rec repair. Studying DNA replication in excision-repair deficient *E. coli* cells, they found that after UV-irradiation the newly synthesized polynucleotide sequences were much shorter than without irradiation. But with the passage of time the difference disappeared, as shorter sequences were apparently converted to (or integrated into) those of normal length. Estimates indicated that the average length of first-synthesized DNA sequences roughly matched the average distance between dimers in DNA of the irradiated cells. The interpretation that blockage of DNA synthesis by a pyrimidine dimer in the parental strand causes formation of a "gap" in the daughter strand is supported by other evidence, for example, that partial photoenzymatic repair of pyrimidine dimers prior to DNA synthesis leads to correspondly longer newly synthesized sequences.

It was further observed that DNA synthesized by UV-irradiated *recA* cells remains in relatively short sequences, even after prolonged incubation, in contrast to wildtype, *recB*, or *recC* cells. This suggested that *closing the gaps* is a physicochemically recognizable reaction step characteristic of this repair process. The length of gaps in newly synthesized strands is of the order of a thousand nucleotides, exceeding by far the lengths of sequences resynthesized in excision repair.

Most likely several pathways for recombination repair exist, all controlled by the *recA* gene. They involve various other *rec* genes and the *lexA* (or *exrA*) gene, but their differences and characteristics require further clarification. This picture is consistent with the findings that *recB*, *recC*, or *lexA* cells are by far not as sensitive as *recA* cells, and that *recA recB* double mutants are no more UV-sensitive than *recA* mutants. More details about the present state of research in this area can be obtained from an overview by P. Howard-Flanders (1975).

The close relationship of this type of repair with genetic recombination is evident from the fact that many of the genes involved in genetic recombination likewise affect the UV sensitivity of the cells, though rarely in such a drastic way as the *recA* gene. Extensive work by Rupp and coworkers showed that in UV-irradiated *uvr⁻ rec⁺* cells DNA strands newly synthesized during replication contain interspersed sequences of parental DNA. This is due to the frequent occurrence of recombination events between sister duplexes; quantitative evaluation of the data suggests one exchange for every one to two pyrimidine dimers, but the specific mechanism for such exchanges is not fully clarified. The schematic representation in Figure 7.12, which indicates exchanges between two sister strands (rather than between all four) of the two sister duplexes is in best agreement with the presently available data.

7.6.2 Complementation between recombination repair and excision repair

Comparison of the survival of wildtype cells with the survival of either *recA* or *uvrA* cells demonstrates that neither one of the two fundamental dark repair systems achieves anything close to what their combination does (see

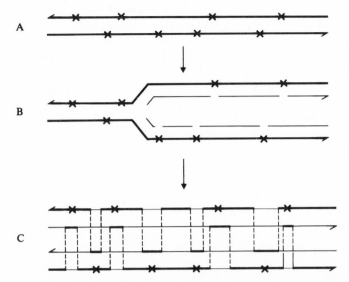

Figure 7.12. Schematic representation of sequential steps in recombination repair of *E. coli*. (A) UV-irradiated double-stranded DNA containing pyrimidine dimers (×); (B) the same DNA during replication, with daughter strands (thin lines) showing gaps where the parental template contains a dimer; (C) filling the gaps with nucleotide sequences from the homologous parental strand by recombination. (Modified from W. D. Rupp et al., *J. Mol. Biol. 61*, 25, 1971.)

Figure 4.7). This shows that the two repair systems complement each other, as one might well expect from their different mechanisms. Moreover, the resemblance of the general UV-sensitivity level displayed by the survival curves of *uvrA* and *recA* cells could suggest similar effectiveness of the two repair systems. However, we should be aware of the fact that survival curves of repair-deficient strains necessarily cover a rather limited fluence range, corresponding to relatively low numbers of UV lesions per genome. This can be grossly misleading for an evaluation of relative merits of the two repair systems when operating in irradiated *wildtype* cells.

For a discussion of this matter we have to remind ourselves that the UV fluences applied to wildtype cells are usually 20 to 50 times higher than those applied to *uvrA* or *recA* cells. Because the two repair systems act sequentially, it is inevitable that the repair system operating first must abolish more than 95 percent of the total UV lesions in order to account for the high wildtype survival, relative to that of a single-repair-deficient strain. Thus the excision repair system, which operates essentially prior to replication, is the one that eliminates the vast majority of the UV damage in wildtype cells.

But even after this extensive excision repair, UV-irradiated wildtype cells may still carry a hundred or more potentially lethal lesions per genome. None of the cells would survive unless the remainder of the lesions were taken care of by recombination repair. This explains why the survival of *recA* cells (displaying excision repair only) is still rather low. Of course, the important but still unanswered question is *why* does the excision repair system leave 3 to 5 percent of the lesions unrepaired while it eliminates all of the others. Possible answers would be: (1) insufficient time for repair; (2) irreparability of certain photoproducts; or (3) occurrence of additional alterations at the sites of excisable lesions, which render them irreparable.

The sequential operation of excision repair and Rec repair implies termination of the former, when new DNA is synthesized with gaps at the location of dimers in the parental strand. Such termination may be automatic because of the requirement of an intact complementary strand for the excision process. Without it, double-strand breakage of the DNA molecule and lethality would be the likely result. Perhaps after the gap closing further pyrimidine dimers are eliminated by a second round of excision repair.

For every UV lesion repaired, recombination repair involves a much longer nucleotide sequence than excision repair. Therefore, recombination repair can only be highly effective (in terms of *percentage* of lesions repaired) when it is confronted with relatively few UV lesions. Whereas this is generally the case at low fluences, it is the case at high fluences only if the vast majority of the lesions are previously excision-repaired. A comparison of the survival curves of *recA* cells and wildtype cells supports this view (see Figure 4.7). The initial portions of the curves diverge most, while their final slopes (if they were plotted on an identical fluence scale) differ by less than a factor of 2.

Quite the opposite is true for excision repair. *recA* cells, where excision repair is the only dark repair mechanism, typicallly show upwardly concave survival curves, suggesting that the percentage of excision-repaired lesions *increases* with the UV fluence. Such curves are also observed with *E. coli* B (see Figure 4.11), which in several respects resembles *phenotypically* a *rec⁻* strain, and with phage T1, T3, and T7, whose host-cell reactivation is independent of the bacterial recombination repair system (see Figure 4.10). Phage lambda, whose host-cell reactivation involves bacterial Rec repair, shows a convex survival curve, with the notable exception of infection of *recA* host cells (Figure 7.20).

7.7 Multiplicity reactivation of viruses

Inactivation of a phage is defined as loss of its ability to produce viable progeny in the bacterial host. However, because in the case of UV inactivation the entire DNA molecule may contain only a few photochemically altered nucleotides and large regions are free of damage, inactivated phages are still capable of carrying out many of their viral functions. In addition, *multiple infection* of host cells can lead to phage survival much higher than expected from the individual survival chances of singly infecting particles. This phenomenon was first recognized by Luria in 1947 and was called multiplicity reactivation.

Let, for simplicity, the surviving fraction of singly infecting phage be approximated by a one-hit, one-target function $S/S_0 = e^{-cF}$. Production of viable phage progeny by cells infected with exactly n phages should then follow the n-target curve $S/S_0 = 1 - (1 - e^{-cF})^n$, if there were no interaction of the inactivated phages with each other. Thus no point of the survival curve should be more than a factor n higher than the survival curve of singly infecting phage. The much enhanced phage survival in multiply infected cells observed by Luria was, therefore, interpreted as the result of cooperation between two or more inactivated phages for survival.

Multiplicity reactivation has been most thoroughly investigated for the T-even phages, where the effects are very extensive, as the curves in Figure 7.13 show. Less extensive multiplicity reactivation occurs in most, if not all, phages containing double-stranded DNA; the effect is often enhanced in the absence of host-cell reactivation.

The extent of multiplicity reactivation depends on the *average multiplicity of infection*, which is defined as the ratio [number of infecting phages] : [number of host cells] in the reaction volume. Because infection results from random interaction between phage and bacterial cell wall (the latter containing many phage receptor sites), distribution of phages among *individual* infected cells can be approximated by *Poisson* terms. Accordingly, the fraction f_n of cells infected with exactly n phages (where $n = 0, 1, 2, 3, \ldots$), when the *average* multiplicity of infection is m, is expressed by

$$f_n = \frac{m^n \cdot e^{-m}}{n!} \tag{7.5}$$

Consequently, the fraction $f_0 = e^{-m}$ of a bacterial cell population is uninfected, whereas the complementary fraction $(1 - e^{-m})$ is infected. Among the infected cells, the subfraction of those singly infected (*monocomplexes*) equals $me^{-m}/(1 - e^{-m})$, whereas those multiply infected (*multicomplexes*) equal $1 - [me^{-m}/(1 - e^{-m})]$.

The essential difference in survival exists between mono- and multicomplexes; therefore, as a first approximation, the theoretically desirable distinction between multicomplexes consisting of 2, 3, 4, . . . infecting phages can be neglected. Typically, we have to consider two experimental situations: (a) *multiple-infection conditions*, corresponding to high multiplicity of infection (for example, $m > 4$), where the great majority of cells are infected by two or more phages; and (b) *single-infection conditions*, corresponding to low multiplicity of infection (for example, $m < 10^{-2}$), where most of the cells remain uninfected, and the few infected cells have seldom taken up more than one phage. At intermediate multiplicities (say, between 0.2 and

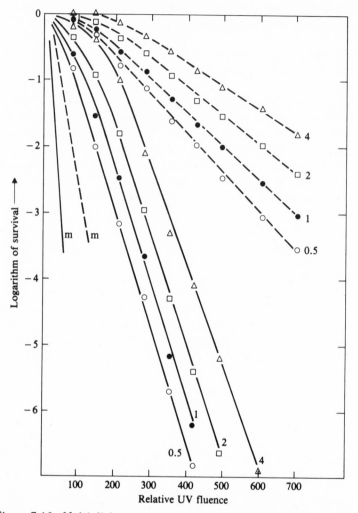

Figure 7.13. Multiplicity reactivation of UV-irradiated phage T2. Solid lines refer to the dark survival of monocomplexes (m) and of multicomplexes (numbers referring to the multiplicity of infection). Dashed lines represent the analogous survival curves after maximal photoreactivation. (From R. Dulbecco, *J. Bacteriol. 63*, 199, 1952.)

2) there is an appreciable mixture of singly and multiply infected cells, which is less suitable for a quantitative study of multiplicity reactivation. To determine the multicomplex survival under these conditions, the experimental data must be corrected graphically or arithmetically for the fraction of mono-complexes and their known survival function.

Even at the lowest feasible multiplicities, where the great majority of infected cells are monocomplexes, multiplicity reactivation is evident. At low survival levels, the inevitable small fraction of multicomplexes[1] causes a significant deviation from the extrapolated monocomplex curve, which finally leads to a slope characteristic of complexes containing two phages. For example, at a multiplicity of 10^{-2} the deviation becomes apparent at survival levels below 10^{-2} to 10^{-3}. Even plating of the unadsorbed irradiated phages with indicator bacteria results in a small fraction of multiply infected cells, which exhibit multiplicity reactivation. Thus it is practically impossible to obtain a pure monocomplex curve to survival levels of 10^{-5} to 10^{-6} for phages containing double-stranded DNA. In contrast, strictly exponential functions to the 10^{-6} survival level and below are observed for the single-stranded DNA phage ΦX 174, which lacks the capability of multiplicity reactivation.

7.7.1 Mechanism of multiplicity reactivation

In essence, multicomplex survival curves differ from monocomplex curves in two respects: They show a considerable shoulder at low UV fluences, and their final slope at high fluences is much reduced. These characteristics presumably reflect the effectiveness of cooperation in different fluence ranges.

Genetic recombination. Luria originally proposed that multiplicity reactivation results from recombination of undamaged parts of different phages infecting the same cell. This idea was abandoned several years later when quantitative predictions of the recombination hypothesis were not met by subsequent experimental results. However, as it turned out, this "failure" was due to invalid specific assumptions in the mathematical formulation of the hypothesis, rather than incorrectness of the proposed basic mechanism. The capability of UV-inactivated phages to undergo genetic recombination is evident from the phenomenon of *marker rescue* (Section 7.8), and involvement of recombination in multiplicity reactivation has afterwards been explicitly shown in a number of cases. There is little doubt today that genetic recombination is the essential, but in some cases perhaps not the only, mechanism accounting for multiplicity reactivation.

An important step toward this end was an investigation by R. H. Epstein in 1958, concerning the composition of *single bursts*[2] from cells multiply infected with UV-irradiated T4 phages differing in two genetic markers. The progeny, being essentially the product of multiplicity reactivation between parental particles, contained a high fraction of recombinants. In particular, the two reciprocal recombinant types were found in highly disproportionate numbers, some bursts containing between 80 and 100 percent recombinants of *one* type only. These results, which are quite dissimilar from those obtained with unirradiated phages, are easily understood on the assumption that for multiply infecting, UV-damaged phages genetic recombination is a *prerequisite* for the production of viable progeny.

Similarly, these findings suggest that the increased recombination frequency observed with UV-irradiated phages (Section 10.5) is the result of multiplicity reactivation, which is inevitably included in such experiments. It is not difficult to conceive that genetic recombination between multiply infecting, genetically marked phages is a dispensable interaction when they are *unirradiated*, but that the interaction becomes indispensable when they are irradiated and must undergo multiplicity reactivation in order to give rise to progeny. Thus multiplicity reactivation constitutes a *selection* factor in favor of genetic recombination in general, and therefore also in favor of the particular recombinant type(s) scored.

Convincing evidence that in phage *lambda* genetic recombination is the only mechanism responsible for multiplicity reactivation is provided by results of R. J. Huskey. The lambda hybrids employed in these experiments were either proficient (red^+) or deficient (red^-) for recombination of vegetative phage particles, and their multiplicity reactivation was studied in $uvrA$ host cells (permitting extensive multiplicity reactivation) that were either recombination-proficient ($recA^+$) or recombination-deficient ($recA^-$). The data, plotted in Figure 7.14, show most extensive multiplicity reactivation effects if both the

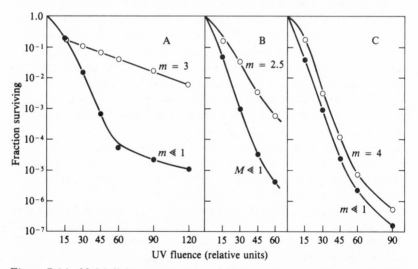

Figure 7.14. Multiplicity reactivation of UV-irradiated lambda hybrids. The survival of single infecting (•) and multiply infecting (○) phages is shown as a function of relative UV fluence. *Panel A: red$^+$* phages infecting *recA$^+$uvrA$^-$* cells. *Panel B: red$^+$* phages infecting *recA$^-$uvrA$^-$* cells. *Panel C: red$^-$* phages infecting *recA$^-$uvrA$^-$* cells. (Drawn after tabulated data by R. J. Huskey, *Science 164*, 319, 1969.)

bacterial *and* the phage recombination systems are functioning (panel A). If only the phage recombination system is operative, multiplicity reactivation is much reduced (panel B), and in the absence of both recombination systems no multiplicity reactivation is observed (panel C). In the latter case, the survival curves of multicomplexes and monocomplexes differ at all fluences only by a factor approximating the multiplicity of infection.

Functional cooperation. Another possible mechanism of multiplicity reactivation, which can be postulated on theoretical grounds, is *functional cooperation*. It is known that phages confined within the same host cell can cooperate with one another by complementing necessary gene functions, as was first shown by Benzer for T4*r*IIA and T4*r*IIB mutants. Neither of the mutants alone is capable of reproduction in *E. coli* K12(λ) cells, but progeny is formed as a result of mixed infection because either one of the mutants contributes the gene function missing in the other. Such complementation occurs even if one of the two mutant types is UV-inactivated, provided the gene itself is still functioning. This is likely because a specific gene function represents a much smaller UV target than the whole phage as plaque former.

It seems at least theoretically feasible that multiplicity reactivation in the T-even phages, in contrast to lambda, is partially a result of functional cooperation. Such an explanation would be compatible with the existence of genes with vital prereplication functions, which occupy a quarter or a third of the genetic map of the T-even phages. In particular, genes involved in phage-specific repair processes (see Sections 7.5 and 7.10) should be able, by functional cooperation, to enhance the survival. UV lesions in such genes may, but need not, be of the solely functional type (Section 6.3).

7.7.2 Significance of multiplicity reactivation

Although multiplicity reactivation has been mainly investigated after UV irradiation, it occurs as well after various other kinds of DNA damage. This is expected for a process of a cooperational nature because realignment of *un*damaged genome regions, or cooperation between *un*damaged functions, should not depend on the kind of damaging agent. Of course, the phage DNA in its entirety must be able to enter the cell. This is hardly a problem after UV irradiation, except at extremely high fluences or wavelengths outside the 240-280 nm region. However, after exposure to ionizing radiations, nitrous acid, formaldehyde, or other chemicals, the phage DNA has greater difficulties in reaching the interior of the host cell, and multiplicity reactivation is much reduced. Significantly, if ionizing radiation is applied *after* infection, multiplicity reactivation is as extensive as after UV irradiation.

Multiplicity reactivation seems to be characteristic of all phages undergoing genetic recombination. The effect is very small or absent in single-stranded DNA phages like ΦX 174 and S13 and in RNA phages, a finding that tends to confirm the rule, as these phages show a very low frequency of genetic recombination. Multiplicity reactivation is also found in animal viruses containing double-stranded DNA, and a small effect has been reported for type 1 *polio virus*, which contains RNA. It is possible, but has not been explicitly shown, that processes analogous to multiplicity reactivation occur generally in cellular organisms containing more than one genome copy per cell or per nucleus. This is often the case during or after DNA replication even in so-called haploid organisms like bacteria or haploid yeasts. Presumably the basic processes would resemble those involved in postreplication repair.

7.8 Marker rescue in phages

When a host cell is infected with a UV-inactivated phage and one or several unirradiated phages, some of the progeny particles may carry genetic markers originating from the inactivated parent. This phenomenon, called marker rescue or cross-reactivation, was extensively studied in the 1950s and 1960s by A. H. Doermann and collaborators. Evidently, the marked DNA region of the inactivated phage is integrated into DNA of unirradiated phage or their descendants as a result of genetic recombination. If the parent was heavily UV-irradiated, the progeny particle carrying the marker derives most of its genome from the unirradiated parent and can hardly be regarded as reactivated from a previous inactivation. Therefore, the term marker rescue describes the phenomenon better than the term cross-reactivation and should be preferred.

Like multiplicity reactivation, marker rescue is not specific for UV lesions because of the underlying recombination mechanism. Marker rescue is found in phage inactivated by decay of radioactive phosphorus (^{32}P) incorporated in its DNA, or in phage inactivated by ionizing radiation, nitrous acid, formaldehyde, or other agents. The essential requirement for its occurrence is that the damage occurs in DNA, which must still be able to enter the host cell. This requirement is fully met by UV radiation from 240 to 280 nm, but only partly by other UV wavelengths, ionizing radiations, or various chemical agents. Less extensive marker rescue after UV irradiation at 285 to 290 nm, or below 240 nm, is probably the result of additional protein damage affecting the phage infection, which makes the analysis of data obtained with 254 nm radiation particularly valuable.

Phage progeny from individual host cells have been tested by the *single-burst technique* for the proportion of phage carrying particular markers from the irradiated parent. This rather laborious work has led to important conclusions regarding the mechanism of marker rescue because it permits us to follow the fate of *several* markers within any particular host cell. The results can be summarized as follows.

If the irradiated and unirradiated parent differ in three unlinked markers, the seven possible genotypes that involve one or several markers from the irradiated parent vary widely in their frequencies *within individual bursts* and also *between different bursts*. With increasing UV fluence, the average number per burst of progeny particles carrying a specific marker of the irradiated parent decreases (although the total burst size remains essentially constant), until particular markers are completely absent (marker knockout). This permits an estimate of the fluence function of marker survival. For the following consideration let $p(F)$ be the probability for survival of a specific marker (*i.e.*, for its reappearance in *at least* one viable progeny particle of the complex), and $1 - p(F)$ be the probability for marker knockout. In the case of three unlinked markers we then expect the various possible bursts to have the following relative frequencies: bursts showing survival of all three markers = p^3; bursts in which any two of the three markers survive = $3p^2 (1 - p)$; bursts in which any one of the the three markers survives = $3p (1 - p)^2$; and bursts with none of the markers surviving = $(1 - p)^3$. From results obtained with phage T4 at various low UV fluences Doermann and collaborators calculated that p decreases exponentially with F at a rate approximating 4 percent of the rate at which the phage loses its plaque-forming ability.

Later work on marker rescue was greatly facilitated by the use of selective techniques, based on the ability of wildtype T4, in contrast to T4rII mutants, to propagate in K12(λ) cells. Mixed infection of *E. coli* B cells with UV-irradiated T4r^+ at low multiplicity and unirradiated T4rII at high multiplicity, and plating of the complexes on K12(λ) indicator, results in plaque formation only if at least one viable T4r^+ particle emerges from the complex. This method permits detection of one or a few plaque-forming complexes among a total of 10^4 or more, so that the survival of the r^+ marker can be followed down to very low levels.

Typical examples for marker rescue of rII$^+$ markers as a function of UV irradiation with up to 1000 mean phage lethal fluences are shown in Figure 7.15A. The slopes of the curves decrease gradually, but become constant beyond approximately 300 mean lethal fluences. For a single marker the constant slope is only about $\frac{1}{150}$ of that found for the T4 survival, and straight extrapolation to zero fluence intercepts the ordinate axis below 10^{-2}. These curve characteristics are observed irrespective of the particular rII marker used. Rescue of closely linked marker pairs occurs at frequencies moderately lower than observed for single markers. Rescue of the pair is a single event: This is evident from the result that after introduction of a third marker between the two existing ones the observed three-marker rescue curve is identical with the rescue curve for the two outside markers alone (not shown in the graph). Theoretical considerations lead us to the same conclusion: If rescue of a single marker occurs with a probability of 10^{-3}, the probability for two *independent* specific rescue events occurring within the same host cell should be 10^{-6}, and the chance for their showing up in one and the same progeny particle would still be considerably lower. This is in clear disagreement with the data.

We notice in Figure 7.15A that the slopes of curves for various marker pairs

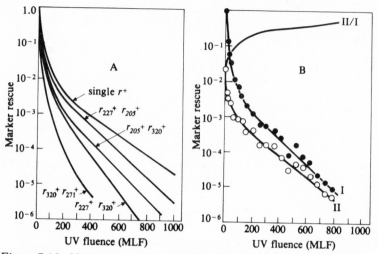

Figure 7.15. Marker rescue in irradiated phage T4. *Panel A*: Survival of a single *r*II$^+$ marker (uppermost curve) is compared with the survival of four *r*II$^+$ marker pairs, as a function of UV fluence (expressed in units of mean lethal fluence [MLF], or approximately 5 J · m^{-2}). The slopes of the survival curves increase as the distance between the two markers in a pair increases. *Panel B*: The rescue of the *r*145$^+$ marker in a wildtype *r*II region (curve I) is compared with the rescue of the *r*145$^+$ marker surrounded by markers *r*168 and *r*147 (curve II). The ratio between the two curves (II/I) indicates the extent at which the rescue of the *r*145$^+$ marker is hampered by the surrounding *r*II markers. (From A. H. Doermann, *J. Cell. Comp. Physiol. 58*, Suppl. *1*, 79, 1961.)

differ; their steepness increases with the distance between the markers. Doermann also observed that an "extended" single *r*II$^+$ marker, corresponding to a short DNA *deletion* in the rescuing phage, is inactivated at a rate typical of two markers whose distance corresponds to the length of the deletion. These results are well understood on the basis of the following model.

A formal model for marker rescue. As pointed out in Section 5.1, the processes leading to marker rescue in phages closely resemble those occurring in bacterial transformation with UV-irradiated DNA. UV lesions have fixed positions in the irradiated phage genome; marker rescue requires that the marker and an undamaged DNA region to the right and left of it be integrated into a viable genome that becomes a mature phage. The marker survival decreases with increasing fluence, as the size of the *undamaged* region carrying the marker (and thus the chance for its rescue) becomes smaller. The expected consequences are those actually observed: (1) The number of complexes exhibiting marker rescue decreases with the fluence, and (2) within the complexes displaying marker rescue, the average clone size of marker-carrying phages decreases because at a decreased recombination probability the recombination events occur, on the average, at a later intracellular state.

A quantitative expression for the expected marker rescue, derived from this model by Meselson and Stahl, is the inverse-square function (equation 5.1) discussed in the context of bacterial transformation. At high UV fluences the experimental curves in Figure 7.15A differ from the inverse-square function in that they decrease linearly in a semilog plot, rather than with continually decreasing slope. This relationship indicates that a small target of fixed size must remain unhit, or else marker rescue fails. As far as rescue of a marker pair is concerned, the obvious explanation is that the DNA region between the two markers must remain undamaged; but this would not hold for single markers, which typically differ from the rescuing genome in only one nucleotide pair. In these cases we may assume that either the altered DNA structure adjacent to a pyrimidine dimer prohibits recombination in this region, or the high UV fluence causes, with a small probability, additional damage interfering with a necessary step in infection as, for example, phage adsorption or DNA injection.

Further support for the model was obtained by Doermann in experiments where the rII^+ marker of the irradiated phage was flanked by rII markers. In this case, only rescue of rII^+ coupled with *nonrescue* of both rII markers can

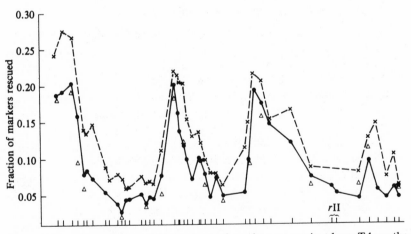

Figure 7.16. Dependence of the extent of marker rescue in phage T4 on the chromosome map location. Experimental data show the survival of two rII^+ markers and 43 am^+ markers located at the relative map positions shown on the abscissa. The latter represents more than three-quarters of the T4 chromosome, i.e., the region from gene 1 to gene 47. Phages were irradiated at fluences corresponding to either 35 (dashed line) or 51 (solid line) phage lethal hits. Triangles correspond to the same phages irradiated with 51 phage lethal hits but infecting another host cell strain. (From F. C. Womack, *Virology* 26, 758, 1965.)

lead to plaque formation of the complexes on K12(λ) indicator cells. In accordance with expectation, this restriction decreases the likelihood of plaque formation at all UV fluences (including zero fluence), as seen in Figure 7.15B. Furthermore, the curves for restricted and unrestricted rescue converge, as the model predicts, because with increasing UV fluences the lesions closest to the r^+ marker successively take over the role of the flanking rII markers in limiting the rescuable region.

This model, based on results obtained with rII markers of T4, probably holds in its general form for other markers as well, except that rescue probabilities may differ. This is evident from marker rescue studies with T4 *amber* mutations, located in many different regions of the chromosome. Since *am* mutants cannot propagate in wildtype *E. coli* cells but in cells carrying an *amber* suppressor (for example, *E. coli* CR63), selective methods analogous to those used for rII markers can be applied. As Figure 7.16 indicates, the T4 chromosome consists of at least three regions of high marker rescue activity, separated by regions of low marker rescue activity; the rII genes are actually located in one of the latter. The differences shown in the graph are independent of the inactivating wavelength, suggesting that recombination events, rather than UV lethal lesions, occur at greatly varying frequencies in different regions of the T4 chromosome.

7.9 Weigle recovery (or UV reactivation) of viruses

In 1953, J. J. Weigle made the paradoxical observation that the survival of UV-irradiated phage lambda can be enhanced by additional UV exposure after infection of the host cell. He also noted a similar effect on survival when the irradiated phage infected host cells *previously* irradiated with UV fluences of the order of 10 to 100 J · m^{-2}. This phenomenon was called *UV restoration* or *UV reactivation*, but in the recent literature the term *Weigle recovery* has often been used, which seems preferable for several reasons. It makes confusion with other recovery effects after UV irradiation unlikely, and discourages the incorrect association of the abbreviation UVR (for UV reactivation) with the gene symbol *uvr*. At the same time the new term permits the inclusion of similar phage recovery effects observed when host cells were pretreated with ionizing radiation or with certain chemicals.

Weigle recovery is commonly found in phage, though often to a lesser extent than in lambda. It is absent in typical virulent phages like the T-even group and T5, but has been observed in animal viruses, for example, in *Herpes simplex* virus infecting monkey kidney cells.

Theoretical evaluation of experimental data on Weigle recovery often suffers from the manner in which they are obtained and presented. A plot like that in Figure 7.17 demonstrates very well the presence or absence of the phenomenon, but a *quantitative* assessment of the underlying repair is

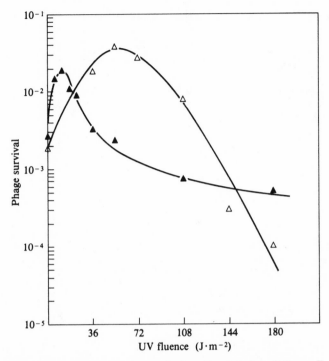

Figure 7.17. Weigle recovery of phage lambda in wildtype *E. coli* C cells (△) and in excision-repair deficient mutant cells C *syn*⁻ (▲). The phage was irradiated with 216 J · m⁻² (△) or 30 J · m⁻² (▲) to yield an approximately equal survival percentage (shown on the ordinate) in unirradiated cells. The abscissa represents the fluence with which the cells were irradiated prior to phage adsorption. (From W. Harm, *Z. Vererbungslehre 94*, 67, 1963.)

rather difficult owing to the superposition of an antagonistic effect, namely inhibition of host cell reactivation. In contrast, if the phage survival is plotted as a function of UV fluence, with different UV exposures of the host cells as parameters, the two effects can be separated. Figure 7.18 shows such data obtained with the *Serratia* phage sigma; the slopes of curves obtained with preirradiated cells are only about 0.25 of the slope obtained with un-irradiated host cells, indicating a repairable sector of 0.75. While this value remains constant with increasing UV exposure of host cells, the fraction of cells displaying host cell reactivation becomes more and more reduced. We can conclude, therefore, that Weigle recovery of this phage is due to repair of 75 percent of those lethal lesions left over by excision repair.

Weigle pointed out in his original publication that, under the experimental conditions enhancing phage survival, the frequency of UV-induced mutations

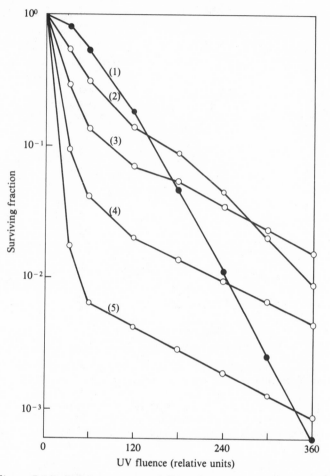

Figure 7.18. Weigle recovery of *Serratia* phage sigma. Phages irradiated with the relative fluences indicated on the abscissa were absorbed to wildtype cells of *Serratia marcescens CV*, which were either unirradiated (curve 1), or pre-irradiated with the relative fluences 60 (curve 2), 120 (curve 3), 180 (curve 4), or 240 (curve 5). (From S. Schulz, doctoral dissertation, University of Frankfurt/M., Germany, 1970.)

among the survivors is likewise increased. This effect, which has been confirmed with other phages, shows that the investigation of Weigle recovery is not only of interest in its own right, but is of potential help in understanding UV-induced mutagenesis. The additional repair in phage DNA, which is found when host cells are either UV-irradiated or otherwise pretreated in a manner affecting their own DNA, is evidently *error-prone*, resulting in occasional mutations. To explain this, we can consider essentially two alternative hypotheses: namely, that either (1) Weigle recovery reflects the *enhancement* of an existing error-prone bacterial repair system; or (2) Weigle recovery is the result of a *newly induced* repair activity.

The fact that both Weigle recovery *and* increased mutability are found in either wildtype or uvr^- cells but are absent in cells lacking either the $recA^+$, exr^+, or lex^+ function favors the first alternative, implying that an enhancement of bacterial recombination repair is involved. Perhaps the repair system can operate more effectively on phage DNA if lesions in bacterial DNA tie up some potentially interfering cellular component(s). Implicating Rec repair definitely should not suggest genetic exchanges between phage and bacterial DNA. This old idea, which was favored for some time after Weigle's discovery, is now ruled out by the fact that neither the loss of the phage recombination function (*red*) nor loss of the phage or bacterial attachment regions affects the Weigle recovery of phage lambda.

The *inducible repair hypothesis* has recently gained support by the observation of new polypeptide bands in gel electrophoresis with UV-irradiated cells proficient for Weigle recovery (see Section 9.4). Whether or not Weigle recovery and UV mutability result from cooperation of such an inducible protein with the $recA^+$ and exr^+ products remains to be seen. Nevertheless, Weigle recovery, which was long considered a specialty of phages and other viruses, has recently attracted considerable interest because it may reveal a mechanism of rather general importance. The inducible repair hypothesis was originally invoked to explain Weigle recovery observed in single-stranded DNA phages. However, because pretreated host cells may have been altered in many ways, one cannot be sure that Weigle recovery of single-stranded DNA phage results from the same processes that lead to recovery of double-stranded DNA phages. If it does, further study of Weigle recovery of single-stranded DNA phages would seem promising because the possibilities for repair of photoproducts in a single DNA strand are rather limited.

7.10 Other phage recovery effects

v-Gene recovery of T4 (Section 7.5), the first discovered reactivation effect requiring genetic information by a phage genome, is the result of excision repair. A search for $T4v^-$ mutants unexpectedly turned up a mutant (x^-) with a UV sensitivity between that of $T4v^+$ and $T4v^-$; its survival curve is seen in Figure 7.19. In combination with the v^- mutation, the mutant is considerably more UV-sensitive than either one of the single mutants, suggesting that x^- and v^- represent deficiencies in two different repair systems.

$T4x^-$ phages show only one-third to one-fifth of the normal frequency of recombination, and marker rescue is considerably reduced when both irradiated and unirradiated parents carry the x^- mutation. It was pointed out by Harm in 1964 that these findings indicate the existence of a repair system involving genetic recombination, an idea later strengthened by the isolation of recom-

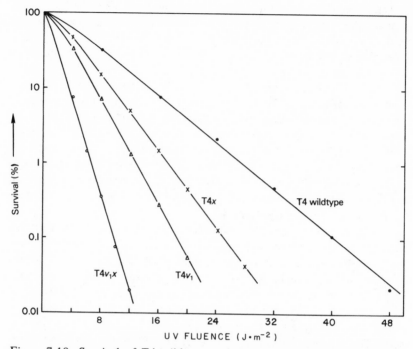

Figure 7.19. Survival of T4 wildtype and the repair-deficient mutants T4x, T4v_1, and T4v_1x as a function of UV fluence. (From W. Harm, *Virology 19*, 66, 1963.)

binationless *E. coli* mutants, which are likewise deficient in repair. The T4 recombination repair system involves at least one other phage gene, y, whose phenotypic effects match those of x, although the two are unlinked.

In agreement with the notion that v-gene recovery and x, y gene recovery are based on fundamentally different repair processes, one finds that UV-irradiated x^- or y^- strains behave like wildtype with regard to pyrimidine dimer excision. Moreover, x^- or y^- mutants, in contrast to T4v^-, have increased sensitivity to ionizing radiations and nitrous acid. Presumably x, y gene recovery occurs in all three T-even phages because T2 and T6 are equally as UV-sensitive as T4v^-x^+, and are more resistant than T4v^-x^-.

A third (and probably last) repair system in T4, which is inhibitable by acriflavine, has been recognized very recently by Van Minderhout and Grimbergen (1975). A mutation in this not yet well-characterized repair system combined with both v^- and x^- (or y^-) mutations further increases the UV sensitivity by a factor 1.3, so that the number of pyrimidine dimers per phage lethal hit is between one and two. This same high level of sensitivity can be observed if v^-x^- mutants infect heavily preirradiated host cells.

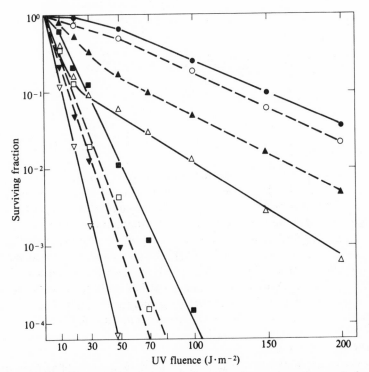

Figure 7.20. Effect of bacterial and phage repair genes on the survival of UV-irradiated phage lambda. λ Wildtype (closed symbols) and λ *red⁻* (open symbols) particles were assayed on *E. coli* K12 wildtype (•, ∘), *recA⁻* (▲, △), *uvrA⁻* (■, □), and *uvrA⁻recA⁻* (▽, ▼) host cells. (From M. Radman et al., *J. Mol. Biol. 49*, 203, 1970.)

Another phage undergoing some kind of recombinational repair after UV irradation is phage lambda. Figure 7.20 shows that the survival of wildtype phage is lower in *recA⁻* than on *recA⁺* cells, indicating that the bacterial recombination repair system abolishes UV lesions in the phage DNA. Although this repair is far less extensive than bacterial excision repair, the two kinds of repair effects are additive. This is evident in *uvr⁺* as well as *uvr⁻* cells, where in both cases the additional *recA⁻* mutation causes a decrease in phage survival. In addition, the lambda gene *red*, which controls recombination of the vegetative phage, plays a role in this or another type of recombination repair. The figure shows that in all cases in which cells are infected by *red⁻* phage, the survival is somewhat lower than when the same cells are infected by wildtype lambda. In consequence, the highest UV sensitivity of phage lambda is observed when *red⁻* phages infect *uvr⁻recA⁻* cells.

8 Repair-related phenomena

8.1 Liquid-holding effects

As early as the middle thirties, A. Hollaender and coworkers realized that UV effects in cells are not determined solely by the amount of radiation energy absorbed. They observed a marked increase in the number of viable cells during lag phase, when an irradiated *E. coli* population, in contrast to an unirradiated control population, was incubated in a poor nutrient liquid medium. The possibility of recovery of inactivated cells was considered at that time, but the problem was not settled. Later, in 1949, Roberts and Aldous reported that the survival of UV-irradiated *E. coli* B can be greatly enhanced by holding the cells for several hours in various nonnutrient liquids. This effect, shown in Figure 8.1, was subsequently termed *liquid-holding recovery,* and strain B has become notorious for its great responsiveness to many kinds of post-UV irradiation treatments.

Irradiated cells of numerous bacterial species show liquid-holding recovery when held in saline or buffered salt solutions. Although it is not obvious that a survival increase resulting from such rather unspecific treatment reflects the same basic processes in different organisms, one can say that at least in *E. coli* it is an expression of enhanced excision repair. Consequently, liquid-holding recovery is absent or negligible in *uvr⁻* strains, but is very extensive in *recA⁻* strains, where essentially all repair is by the excision mechanism. Similarly, phages T1 and T3 show increased survival if phage-infected, excision-repair proficient cells are held in buffer for several hours before plating. This effect is absent if the host cells are excision-repair deficient. Liquid-holding recovery is not limited to prokaryotes; it has also been extensively studied in UV-irradiated yeast, *Saccharomyces cerevisiae.*

The extent of liquid-holding recovery in different excision-repair proficient strains varies greatly, and the effect can be absent altogether. Even a *negative liquid-holding effect* can be observed, by which we mean a survival *decrease* with the time of buffer holding. As shown in Figure 8.2, this is the case in *E. coli* C, and is also observed in the yeast *Schizosaccharomyces pombe.* Both liquid-holding recovery and the negative liquid-holding effect in *E. coli* are strongly inhibited by the presence of caffeine, acriflavine, or potassium cyanide in the liquid medium, three substances known to inhibit excision repair.

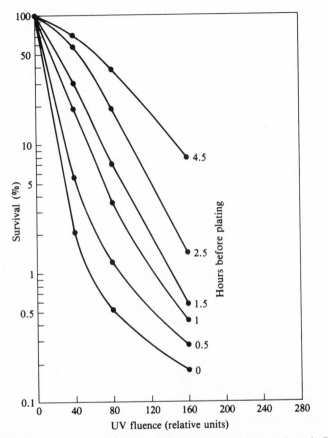

Figure 8.1. Liquid-holding recovery in UV-irradiated *E. coli* B cells. After exposure to various fluences, cells remained suspended in buffer for up to 4.5 hours (as indicated on the curves) before they were plated on Nutrient Broth agar. (From R. B. Roberts and E. Aldous, *J. Bacteriol. 57*, 363, 1949.)

These observations can be accommodated by the general assumption that liquid holding *modifies* the extent of excision repair. Where the extent is relatively low, holding of the cells tends to increase it (resulting in recovery); but where repair is already very extensive without liquid holding, it remains unaffected or even decreases. In this context it is interesting to note that *E. coli* C is more UV-resistant than most wildtype strains of *E. coli,* and that *S. pombe* is a highly UV-resistant yeast species. Why liquid holding increases the cell survival in some cases but decreases it in others, is not yet clear. Obviously, a complex repair process involving several different enzymes requires, for maximal effectiveness, an optimal balance between their reactions. Experimental conditions, of which liquid holding is just one, can be

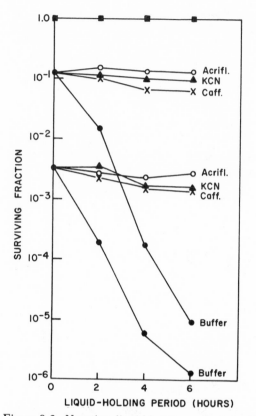

Figure 8.2. Negative liquid-holding effect in UV-irradiated *E. coli* C cells. Cells were irradiated with 80 or 120 $J \cdot m^{-2}$ and kept for up to 6 hours at room temperature in either plain buffer, or buffer containing 10 μg/ml acriflavine, 2 mg/ml caffeine, or 0.002 M KCN, as indicated on the curves. Unirradiated control cells (■) showed no change in viability upon liquid holding. (From W. Harm and K. Haefner, *Photochem. Photobiol. 8*, 179, 1968.)

expected to affect this balance and thus the extent of repair. Such other parameters as *growth medium, temperature,* and so forth, probably do the same. Recent evidence suggests that degradation and resynthesis processes occur regularly even in DNA of unirradiated cells; in irradiated cells such processes may interfere with excision repair to a varying extent, depending on the conditions.

It should also be realized that a small difference in the fraction of lesions excision-repaired may cause a large difference in survival, as the following simple example shows. Exposure of *E. coli* to 2 $J \cdot m^{-2}$ of 254 nm radiation produces approximately 130 pyrimidine dimers per chromosome. In excision-repair proficient *recA*⁻ cells, on the average 125 of them (or 96 percent) are repaired, and a survival of e^{-5}, or approximately 10^{-2}, would be a reasonable expectation. But if the repair were slightly more extensive (repairing an average of 128 dimers out of 130), the expected survival would exceed 10^{-1}, whereas a reduction in repair to 120 dimers would decrease the survival to $<10^{-4}$. In other words: the apparently small difference in the number of repaired lesions between 120 (or 92.5 percent) and 128 (or 98.5 percent) can affect the survival level by more than a factor of 1000.

Experimental results have shown that component steps of excision repair (incision, excision, and some DNA resynthesis) actually occur while the irradiated cells are held in buffer. But evidently the cell survival obtained with such treatment results not only from the repair *during* liquid holding, diminished by possibly *adverse* effects, but also from subsequent repair on the nutrient agar plate. Because at present nothing is known about additivity or overlap of these partial reactions, biochemical or physicochemical measurements of single-strand nicks, dimers excised, or DNA synthesized are not sufficient for understanding the biological result and for making quantitative predictions.

8.2 Fluence fractionation and fluence protraction

In studies of survival kinetics with biological materials the fluence *rate* is often considered insignificant for the result. The reason is that, from a photochemical point of view, the quantity and quality of photoproducts formed should depend solely on the fluence F, that is, on the product of fluence rate with time (Bunsen-Roscoe reciprocity law). However, since UV inactivation of cells is the result of photochemical *and* biological processes, protraction of the exposure over a long time period, or its fractionation into several portions (separated by appreciable time intervals) may affect the survival kinetics under certain circumstances. A rather trivial reason for such an effect would be an alteration in the physiological state of the biological material *during* radiation treatment. For example, if cultured cells are at the beginning in the G1 phase of the mitotic cycle, but reach the S phase before completion of irradiation, their UV sensitivity may have changed.

A dependence of the biological effect on fluence protraction or fractionation can also result from the different distribution of the occurrence of photoproducts *in time*, which can have a considerable effect on the extent of excision repair. Figure 8.3 shows the results of fluence fractionation at different survival levels. After exposure to UV fluences F_1 and F_2, with a fixed liquid-holding period between, cell survival is higher than after a *single* exposure to fluence $F_1 + F_2$ with subsequent liquid holding. The difference between the two conditions is considerably more pronounced if a given fluence is fractionated into four or more portions, each separated by a proportionately shortened liquid-holding period (not shown in the figure).

In the extreme, UV radiation can be applied intermittently, for example by

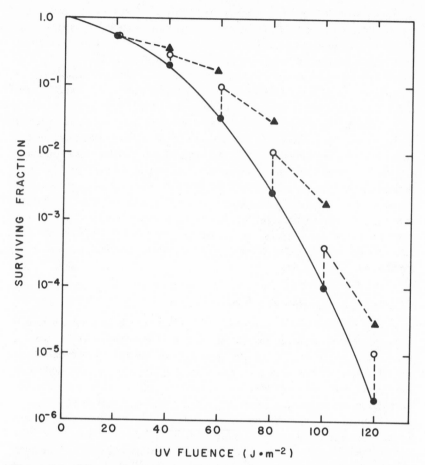

Figure 8.3. Effect of the timing of UV irradiation upon the survival of *E. coli* B/r cells. (•) Cells receiving a single short exposure and plated immediately. (○) Irradiated cells liquid-held for 4 hours at room temperature before plating. (▲) Cells receiving a second exposure (20 J · m^{-2}) *after* liquid-holding for 4 hours. (From W. Harm, *Photochem. Photobiol.* 7, 73, 1968.)

exposing cells through a 1° sector window in a rotating metal disk. At a speed of one rotation per sec, each exposure for 1/360 sec is followed by a dark period of 359/360 sec. As shown in Figure 8.4, this increases the survival of *E. coli* B/r tremendously compared to control cells, which, after irradiation at a high fluence rate, are plated either immediately or after liquid holding for the same period of time required for the fractionated irradiation. Similar results are obtained with *continuous* irradiation at 1/360 the fluence rate applied to control cells.

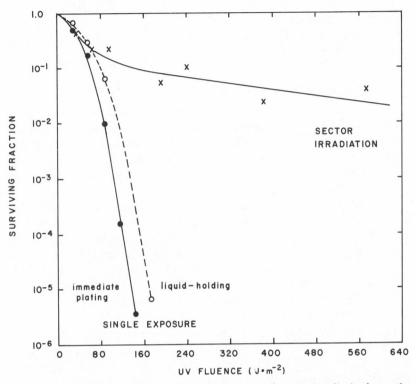

Figure 8.4. Effect of intermittent UV exposure (sector irradiation) on the survival of *E. coli* B/r cells. The fluence rate, averaged over the total duration of the exposure, was 2.2×10^{-3} J $\cdot$ m^{-2} sec^{-1}. For comparison, cells were irradiated in a single exposure at a fluence rate of 0.8 J $\cdot$ m^{-2} $\cdot$ sec^{-1}, and plated immediately ($\bullet$) or after liquid holding (75 min for each 10 J $\cdot$ m^{-2}, in order to match the time required for the intermittent exposure). (From W. Harm, *Photochem. Photobiol. 7*, 73, 1968.)

The impressively high survival can be attributed to very extensive excision repair. Accordingly, *uvr*$^-$ mutants of *E. coli*, or biological systems that cannot be repaired *during* irradiation (for example, extracellular phage, or bacterial transforming DNA), show no greater survival after such fractionated or protracted irradiation than after acute irradiation with the same UV fluence. Why excision repair is so effective under these conditions remains to be clarified. Apparently the important factor is the low *rate* of photoproduct formation in DNA, permitting repair of existing photoproducts concurrently with the occurrence of new ones. As a result, the cellular repair system is at no time confronted with as many UV lesions as are present after irradiation at a high fluence rate. The implication that under these conditions the probability of repair *per lesion* is considerably increased is consistent with R. H.

Haynes's (1966) interpretation of the extensive shoulders characteristic of wildtype *E. coli* survival curves: namely, that they reflect a decreasing repair probability per lesion with increasing fluence, that is, with increasing number of photoproducts present at a time.

8.3 Other factors influencing UV inactivation and cellular repair

Many experimental parameters prior to, during, or after UV irradiation are known to affect the response of biological systems. Most of them influence the physiological state of the organism, in particular the state of its genetic material, which may have two principal kinds of consequences: (1) a change in the quantity and/or quality of photoproducts resulting from a given UV exposure; or (2) an alteration in the extent of repair.

Examples of the first possibility are the formation of bacterial spores and their germination, or irradiation at very low temperatures (see Sections 3.3.3 and 3.3.4). The amount of DNA per cell could likewise be relevant: for instance, fast-growing *E. coli* cells in exponential phase contain two to four genome equivalents, compared with essentially one genome in resting cells, and after UV irradiation the number of photoproducts *per cell* should be correspondingly higher. Nevertheless, because of the redundancy of the two to four genome equivalents, one would naively expect in fast-growing cells a higher survival than in resting cells, but the opposite is often the case.

The likely reason is an alteration in the extent of excision repair, for which less time is available in the growing cells between subsequent rounds of replication. Variations in the extent of repair can also explain differences in the net UV sensitivity of cells that depend on various other experimental parameters. For example, the survival of UV-irradiated *E. coli* B cells is greatly influenced by the postirradiation *plating medium* and *incubation temperature*. A general rule, pointed out by Alper and Gillies, suggests that the survival on various plating media is higher, the slower the growth is on these media. Examples are shown in Figure 8.5. Similarly, increased survival is observed as a result of temporary inhibition of protein synthesis by chloramphenicol.

Incubation of irradiated cells of *E. coli* B at a temperature of 44–45°C, which is just about the upper limit for growth of most *E. coli* strains, leads to maximal survival, whereas the survival is minimal at 25–30°C, as illustrated in Figure 8.6. Transfer of cells from one temperature to another, or from one medium to another, at various times shows that the conditions during the first two to four hours after irradiation are critical for the resulting survival level. The same is found for incubation in an agar medium with a pH of 4.8 to 5.4, where the survival is higher than at neutral or slightly alkaline pH. A great enhancement in survival is also observed in the presence of 0.08 M pantoyl-lactone, a substance counteracting filament formation in *E. coli* B cells.

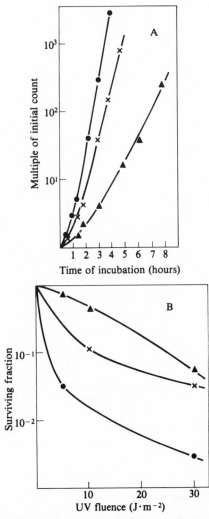

Figure 8.5. Correlation of the rate of growth of *E. coli* B cells in liquid media (panel A) with the UV sensitivity observed after plating on solid media of analogous composition (panel B). (▲) Synthetic minimal medium; (×) Difco Nutrient broth medium (8 g/liter); (•) Oxoid Nutrient broth medium (25g/ liter). (From T. Alper and N. E. Gillies, *J. Gen. Microbiol. 22*, 113, 1960.)

Most of these phenomena reported in the literature during the past 30 years are not understood in detail. As a rule, the effects are small or absent in excision-repair deficient cells, indicating that essentially the extent of excision repair is affected by the various conditions. It is probably not fortuitous that the influence of the growth phase and the incubation medium is most pro-

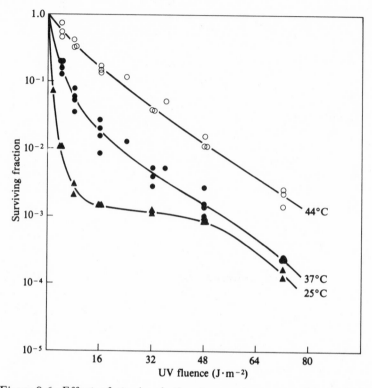

Figure 8.6. Effect of the incubation temperature upon the survival of UV-irradiated *E. coli* B cells plated on Difco Nutrient broth agar. (From W. Harm, unpublished data.)

nounced in *recA⁻* strains and strain B of *E. coli*, where dark repair relies exclusively or mainly on the excision-resynthesis mechanism.

8.4 Photoprotection and indirect photoreactivation

Photoprotection is defined as an increase in survival after far-UV irradiation, resulting from a *preceding* illumination with near UV or visible light. This effect occurs less commonly than enzymatic photoreactivation: It is observed in some *E. coli* strains but not in others, and it has been reported for only a few other bacterial species. The extent of photoprotection in *E. coli* B is shown in Figure 8.7. The action spectrum for causing photoprotection in *E. coli* is narrower than that for photoenzymatic reactivation of the same organism. It has a maximum at 340 nm and declines rapidly at greater wavelengths; those above 400 nm are virtually ineffective.

A survival increase resembling that after preillumination can also be achieved by *post*illumination. Demonstration of this effect requires a photoreactivating-enzyme deficient mutant of a photoprotectable strain, such as *E. coli* B *phr⁻*,

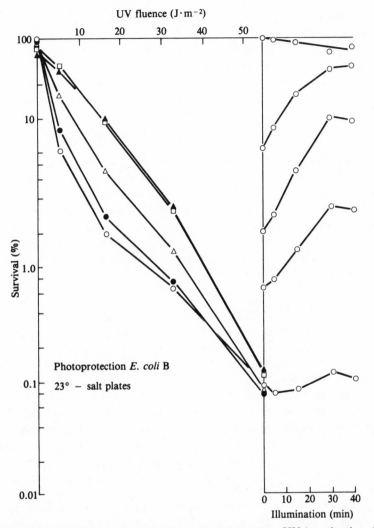

Figure 8.7. Photoprotection of *E. coli* B cells against UV inactivation. The left panel shows the survival of cells kept dark (○), or illuminated with photoprotecting light for 5 min (●), 15 min (△), 30 min (▲), or 40 min (□) prior to UV irradiation. The right panel shows the kinetics of survival increase resulting from preillumination for cells irradiated with various UV fluences. (From J. Jagger, *Radiation Res. 13*, 521, 1960.)

in order to exclude any survival increase due to photoenzymatic repair. Because the postillumination effect fits the operational definition of photoreactivation, it has been called indirect photoreactivation, to distinguish it from ordinary photoreactivation, which involves photoenzymatic repair. Its action spectrum is indistinguishable from that of photoprotection.

Evidently, photoprotection and indirect photoreactivation are two operationally distinguishable phenomena based on the same mechanism. Both occur exclusively in excision-repair proficient strains, and it seems likely that the survival increase reflects an enhancement in the extent of excision repair. The implication that, like liquid-holding recovery, photoprotection and indirect photoreactivation should be observed particularly in strains with less than maximal dark repair, makes its limited occurrence understandable.

The explanation of photoprotection has in the past been closely linked with the induction of *growth delay,* for which the action spectrum is very similar (see Figure 10.6). As discussed below in Section 10.3, the action spectrum expresses near-UV absorption by 4-thiouracil present in some of the transfer RNAs; the resulting photoadduct formation leads to the growth delay. A mutant lacking 4-thiouracil shows neither growth delay nor photoprotection; however, *rel⁻* mutants, which do not undergo growth delay either, nevertheless display photoprotection. Therefore, the long-standing idea that the additional time provided by growth delay enhances excision repair and thus leads to photoprotection seems incorrect; more likely, the 4-thiouracil photoeffect stimulates excision repair in some other manner.

8.5 The Luria-Latarjet effect

In the early years of modern bacteriophage research, T. F. Anderson, one of its originators, suggested determining the kinetics of intracellular phage reproduction by studying the UV survival kinetics of phage-infected cells (complexes). The basic idea was simple: If the survival of extracellular phage follows an approximate one-hit, one-target curve, the survival of complexes should resemble a one-hit, n-target curve (with the same target size as for single phage particles) when the infecting phage had n-fold multiplied. Consequently, determination of the target number n at various times after infection should provide the time function of intracellular phage multiplication.

Experimental work along these lines with phage T2 was carried out by Luria and Latarjet in 1947. However, their results were far from the expectation, a situation not uncommon in experimental science: The curves remained essentially one-target-like, but the slopes decreased with progressing time. This surprising observation, later termed the Luria-Latarjet effect, created new problems instead of solving one already existing.

Subsequent studies of the Luria-Latarjet effect by various scientists, with improved techniques and with different problems in mind, indicated that for a number of reasons the evaluation of such data is generally more difficult than with those obtained with extracellular phages: (1) Intracellular states are not as well defined as the extracellular state of the phage; (2) the synchrony within a population of complexes is not perfect, even with the best techniques available; (3) considerable intracellular DNA replication may cause some self-shielding; (4) because the experiments involve irradiation of

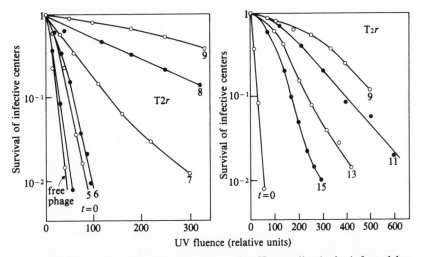

UV fluence (relative units)

Figure 8.8. Luria-Latarjet effect in phage T2. Host cells singly infected by T2r particles were UV-exposed at various times after infection. Times (in minutes) are shown on the curves; the total duration of the latent period was 19 min. The survival curve for extracellularly irradiated T2r is shown for comparison. Notice that the fluence scales differ by a factor of 2. (From S. Benzer, *J. Bacteriol. 63*, 59, 1952.)

the host cells, such effects as Weigle recovery (Section 7.9), inhibition of host-cell reactivation (Section 7.3), or loss of the cells' capacity to propagate phages (Section 10.6) may be superimposed upon the basic phage survival. We will concentrate our discussion on experiments carried out with T-even phages because they are least affected by such effects.

Figure 8.8 shows experiments by S. Benzer on the Luria-Latarjet effect in phage T2, under conditions where the minimum latent period (i.e., the minimum time from infection to cell lysis) is 19 minutes. Three main phases can be distinguished: (1) Through the first 5-6 minutes of intracellular development, before DNA synthesis starts, UV sensitivity decreases slightly; (2) after the onset of DNA synthesis, from the 6th to the 9th minute after infection, UV sensitivity decreases rapidly; and (3) after 9 minutes, when the first mature phage appear within the cell, UV sensitivity increases again. Only in this late period of the intracellular phage cycle do the curves become multitargetlike, as originally expected.

Although not all aspects of the Luria-Latarjet effect are well understood, the present picture can be summarized as follows: The sensitivity decrease during the first 5-6 minutes either reflects the dispensability of certain pre-replication functions, as suggested originally by Benzer, or is the expression of an increasing extent of repair. The former implies the existence of solely functional damage (Section 6.3 and Figure 6.5), which in view of the magnitude of the effect ($\approx 50\%$ of the total phage sensitivity) is alone hardly sufficient. But quite a number of "early" genes are involved in x-gene recovery (Section 7.10), which is common to all T-even phages; and to the extent that their expression occurs prior to irradiation, the total phage sensitivity should decrease.

The strong reduction in UV sensitivity after the onset of DNA replication is best explained by multiplicity reactivation (Section 7.7), occurring as soon as the cell contains more than one phage genome. An illustration of the decrease in relative target size during the first 8 minutes after infection is given in Figure 8.9. It emphasizes the easily overlooked fact that the apparently small sensitivity decrease during the first 6 minutes corresponds to reduction of the original target by approximately 50 percent, whereas the much more impressive further reduction amounts to less. But if the target size *remaining* after the first 6 minutes is considered, the second phase of the Luria-Latarjet effect reduces it by 80–90 percent, which resembles the extent of multiplicity reactivation found with extracellularly irradiated phage.

The third phase in the Luria-Latarjet effect is characterized by the reversing trend in UV sensitivity, which begins at a time when the first phage particles mature within the infected cell. Our understanding of this part of the Luria-Latarjet effect has been greatly improved by the work of N. Symonds. As we see in Figure 8.10, the survival curves become progressively biphasic, that is, the fraction of complexes retaining the high UV resistance decreases, while the remainder show multitarget survival but a much greater sensitivity. Evidently the resistant fraction represents complexes that, as a result of multiplicity reactivation, are still capable of forming at least one viable, mature phage particle prior to cell lysis. The reason for the progressive loss of this fraction is the *predetermination* of the time of lysis at the beginning of the third phase, presumably by the production of lysozyme. Symonds compared the Luria-Latarjet effect of a particular mutant ($T2hs_1r_{34}$) in stationary phase host cells (where lysis is greatly delayed) with that in exponentially growing host cells (which are lysed by this phage at the usual time).

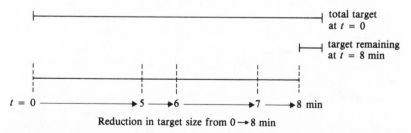

Reduction in target size from $0 \rightarrow 8$ min

Figure 8.9. Schematic representation of the reduction in formal target size during the first 8 min of intracellular development of phage T2r. (Based on Benzer's experimental data shown in Figure 8.8.)

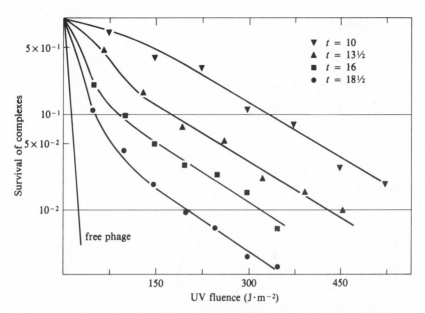

Figure 8.10. Luria-Latarjet effect in phage T2hs_1r_{34}. Complexes irradiated at various times during the second half of the latent period (10, 13.5, 16, and 18.5 min after infection) show heterogenous UV sensitivity. The highest resistance (represented by complexes irradiated at 10 min) is observed in a decreasing fraction of the complexes as time proceeds toward cell lysis, which occurs at 25 min after infection. (From N. Symonds, *Virology 3*, 485, 1957.)

Although in both kinds of cells phage maturation begins at the same time, the high level UV resistance of the phage is retained for a long time in the resting cells, but declines as in Figure 8.10 in the growing cells.

For reasons mentioned previously, studies with other phages are more difficult to interpret. Phages T1 and lambda show extensive Luria-Latarjet effects in excision-repair deficient cells, where host-cell reactivation is absent, but much smaller effects in excision-repair proficient cells. Because the same is true for multiplicity reactivation, we can assume that the latter is responsible for much of the Luria-Latarjet effect in these phages.

Luria-Latarjet experiments with phage ΦX 174 have been carried out with the aim of establishing intracellular changes in the DNA structure of this phage. The important thing to remember is that the infecting particle contains single-stranded DNA, which soon after infection is converted to a double-stranded form (replicative form) with varying structure. In contrast to the infecting single-stranded DNA, the replicative form is host-cell-reactivable; therefore, one would expect soon after infection a more pronounced sensitivity decrease in excision-repair proficient cells than in repair-deficient cells.

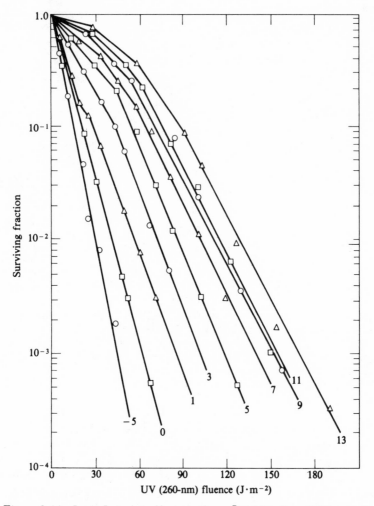

Figure 8.11. Luria-Latarjet effect in phage ΦX 174. *E. coli* C host cells were singly infected and UV-irradiated at the times indicated on the curves. The lowest curve represents the survival of extracellularly irradiated phage. (From D. T. Denhardt and R. L. Sinsheimer, *J. Mol. Biol. 12*, 674, 1965.)

Detailed studies along these lines by Denhardt and Sinsheimer revealed much of the complexity involved in the Luria-Latarjet effect of this phage. At first glance, the results in Figure 8.11 appear to be multitarget curves representing the number of intracellular phages. But actually the extensive shoulders of the curves reflect the rapidly declining capability of the irradiated cells to host-cell-reactivate the replicative form of DNA. Evidence for this is the fact that the limiting slopes in Figure 8.11 resemble those obtained for the replicative form of ΦX 174 DNA in repair-deficient cells. A further complication in the analysis of Luria-Latarjet curves with this particular phage arises from an unusually high UV sensitivity of the host cells regarding their capacity to propagate ΦX 174.

9 UV mutagenesis

9.1 General

The mutagenic action of ionizing radiations was discovered by H. J. Muller in 1927, who later received the Nobel prize. UV radiation was subsequently tested for its possible mutagenicity, and the first positive evidence was obtained by Altenburg in 1930 with *Drosophila*. In the middle and later 1930s studies with similarly favored genetic objects, such as *Zea mays* (corn), *Antirrhinum* (snapdragon), *Sphaerocarpus* (liverwort), *Tradescantia*, and others, confirmed the earlier findings. Although ultraviolet light never matched the importance of ionizing radiation in mutagenesis work with higher plants and animals, the establishment of action spectra led to a most significant result: Prior to any other evidence for DNA being the hereditary material, action spectra showed that UV mutagenesis results from energy absorption in *nucleic acids* rather than proteins. Examples are given in Table 9.1 and Figure 9.1.

As a matter of fact, the strong absorption of UV radiation in biological tissue severely limits its application as a mutagenic agent for higher organisms. To expect reliable results, germ line cells must be either fully exposed to the radiation, or the shielding by other tissue must be slight. Preferred techniques include the irradiation of polar cap cells of the early embryo of *Drosophila*, or the exposure of mature spermatozoa in the superficially located testes of *Drosophila* males. Spermatozoïds of the liverwort *Sphaeroscarpus donnellii* in aqueous suspension are quite suitable, whereas dry pollen grains of angiosperm plants (e.g., *Zea mays*) are less than ideal because of their large diameter (approximately 90 μm). In vitro irradiation of cultured *Tradescantia* pollen tubes, shielded by only a 5-μm layer of cytoplasm, permits immediate cytological investigation of chromosomal alterations.

Compared with most eukaryotic cells, the amounts of UV-absorbing material contained in viruses, bacteria, or some fungi are small. Therefore, in radiation mutagenesis studies with microorganisms ultraviolet light has played the major role. One of the fundamental papers in this field was published in 1946 by Demerec and Latarjet under the title "Mutations in Bacteria Induced by Radiations." Not only did it demonstrate that selective techniques permit accurate determination of UV-induced mutation frequencies within very large populations (in this case: resistance of *E. coli* cells to phage T1), but it revealed a hitherto unknown effect, which remained unexplained for some time: a much delayed appearance of the mutants. The great achievements of microbial

Table 9.1. *Relative effectiveness of various UV wavelengths including mutations in higher organisms*

Wavelength (nm)	Percentage of mutations scored in	
	Sphaerocarpus donellii[a] (liverwort)	*Zea mays*[b] (corn)
248	–	9.9
254	27.8	15.5
265	41.8	11.6
270	–	7.8
280	21.9	9.3
289	–	6.2
297	5.8	4.7
302	5.6	–
313	0.0	–
Unirradiated control	0.0	–

[a]Irradiation of *Sphaerocarpus* spermatozoids in water. Percentages of mutants includes viable and lethal mutations; the latter can be established by tetrad analysis.
[b]Irradiation of *Zea mays* pollen. Th mutations concern deletion mutations leading to loss of various endosperm characteristics.
Sources: (for *Sphaerocarpus donellii*) E. Knapp et al., *Naturwissenschaften 27*, 304, 1939; (for *Zea mays*) L. J. Stadler and F. M. Uber, *Genetics 27*, 24, 1942.

genetics at that time further facilitated UV mutagenesis studies by providing suitable biological systems and assay methods.

The basic concept governing these investigations had been the hope that application of UV radiation would help reveal the properties of the hereditary material and/or the general physicochemical mechanisms underlying mutation. As it happened, however, UV studies contributed relatively little to this general problem; most of our current knowledge was gained by other approaches. Recent experimental work instead addresses specific questions such as these: (1) Which of the primary UV photoproducts in DNA are actually, or potentially mutagenic? (2) Which secondary processes occurring at such photoproducts lead to the stable, replicable DNA alterations that constitute a mutation? (3) Which of the possible types of mutational changes in the DNA molecule can result from the primary UV effects? Answers to these questions, which appear to be less ambitious and less fascinating than those originally envisaged, are nevertheless of considerable interest, as UV radiation is not only a handy mutagen in the laboratory, but also an important environmental factor in the form of sunlight.

Progress toward understanding UV mutagenesis has been made particularly during the past 10 to 15 years in conjunction with studies on repair. The

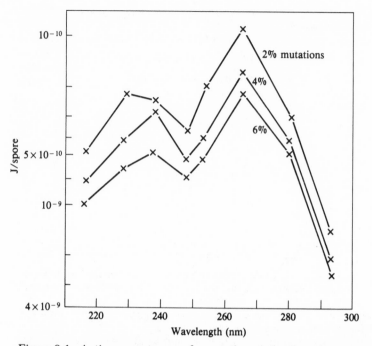

Figure 9.1. Action spectrum of mutation induction in spores of the dermatophytic fungus *Trichophyton mentagrophytes*. The energy required for the production of 2, 4 or 6% frequency of total mutations is shown on an *inverted* logarithmic ordinate scale. The resemblance to the DNA absorption spectrum in the region 235-295 nm is obvious. (From A. Hollaender and C. W. Emmons, *Cold Spring Harbor Symp. Quant. Biol. 9*, 179, 1941.)

following sections will show that, as a rule, the induced mutations do not result from specifically mutagenic photoproducts in the DNA, but are the consequences of errors made in the repair of lethal photoproducts.

9.2 Detection of UV-induced mutations and determination of their frequencies

Because the term mutation is defined as a *hereditary* alteration, its frequency of occurrence as a result of UV irradiation can be quantitatively assessed only for the surviving fraction of a cell population. Therefore, the data are conventionally plotted in the form of *relative mutation frequencies* (i.e., mutants/ survivors) as a function of UV fluence, and it is generally implied that mutations *would* have occurred at the same frequency in the nonsurviving cells. Such an implication is reasonable if mutations are the result of error-prone repair of lethal photoproducts (Section 9.4) because survivors and nonsurvivors presumably differ only by random variations in the actual number of repair events.

Nevertheless, in view of the fact that the nonsurvivors often constitute more than 90 percent of the total irradiated population, it is a legitimate question to ask whether the experimentally observed increase in mutation frequency could not be merely the result of selection in favor of spontaneous mutants already present in the population. However, with the exception of mutations enhancing the radiation resistance itself, there is no support for such an explanation. On the other hand, there is experimental evidence that at least a considerable proportion of the scored mutations must have been UV-induced rather than UV-selected. In many cases this follows from the determination of *absolute* mutation frequencies, that is, the number of mutants among the *total* (viable and inviable) cells. As an example, let us suppose that a certain low UV fluence reduces the cell survival to 50 percent, but causes a fivefold enhancement in the *relative* mutation frequency. In this case the absolute number of viable mutants after irradiation is 2.5 times the number of viable mutants present before irradiation; therefore, *at least* the fraction 1.5/2.5 (or 60 percent) of the scored mutants must have been UV-induced.

Besides the demonstratation that UV causes mutations instead of merely selecting for them, absolute mutation frequencies are of practical concern in attempts at *isolating* particular genotypes. For example, if a population of 10^9 individuals is reduced by the irradiation to 10^5 survivors, it gives no comfort to know that the applied fluence has increased the relative mutation frequency 100-fold from a spontaneous level of 10^{-8}. Most likely, no viable mutant at all would in this case be among the survivors. Because absolute mutation frequencies usually increase at low fluence and fall below the spontaneous frequency at higher fluences, determination of the maximum of the curve is sometimes desirable.

Because absolute and relative mutation frequencies are interrelated by the survival function, they can be calculated from the same experimental data if the survival function is known. However, the original hope that the *shape* of mutation induction curves might enable us to draw direct conclusions concerning the mechanism of UV mutagenesis has diminished. At present, the problem is rather to reconcile the shapes of experimental curves with current knowledge about UV mutagenesis obtained by other approaches. These approaches include the comparison of quantitative data for genetically different strains and for different experimental conditions, and will be presented in the next sections. The kinetics of UV mutagenesis with UV fluence will be discussed in Section 9.5.

9.3 Evidence for a premutational state

9.3.1 Photoreversal of mutations

Like the quantity of lethal events, the frequency of UV-induced mutations is not unambiguously established by the irradiation itself, but depends on the

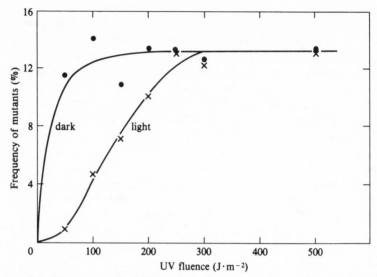

Figure 9.2. Photoreversal of UV-induced mutations in *E. coli*. (Plotted after data by H. B. Newcombe and H. A. Whitehead, *J. Bacteriol. 61*, 243, 1951.)

subsequent treatment. When Kelner discovered photoreactivation in 1949 (see Section 7.2), he also reported a decrease in the frequency of UV-induced mutations in the irradiated cells as a result of illumination with visible or near-UV light. This effect then became known as *photoreversal of mutations*.

Evelyn Witkin (1963) observed extensive photoreversal of certain mutations in wildtype *E. coli*, but no such effect for the same mutations in a *phr⁻* derivative strain. However, for other mutations the photoreversal was equally extensive in *phr⁺* and *phr⁻* cells. This suggested that in some cases photoenzymatic repair is solely responsible for the photoreversal of UV-induced mutations, whereas in other cases the mechanism is of a different kind. Support for the latter hypothesis was obtained by Kondo and Jagger (1966), who showed virtually identical photoreversal of UV-induced prototrophy mutations in *phr⁺* and *phr⁻* cells by 313-nm and 334-nm light, but large differences at greater wavelengths. For the *phr⁺* strain the action spectrum resembles that for photoenzymatic repair, whereas for the *phr⁻* strain it resembles that for indirect photoreactivation or photoprotection. The fact that *phr⁻ uvr⁻* cells show no photoreversal of mutations at any wavelength, whereas the corresponding *phr⁺ uvr⁻* cells show the effect at all wavelengths between 313 and 436 nm indicates that the photoreversal in *phr⁻* cells is due to light-stimulated excision repair.

These results tell us that UV mutagenesis must be largely the result of pyrimidine dimer formation in DNA, the number of dimers being strongly reduced either by photoenzymatic monomerization or by excision in excess of that occurring in the dark. In both cases fewer dimers are retained in DNA for subsequent repair by error-prone, that is, potentially mutagenic, processes. However, some types of mutations are nonphotoreversible; they must be caused by photoproducts other than pyrimidine dimers.

Constant fluence modification factors (as in the case of photoenzymatic repair) are observed for the photoreversal of some mutations but not of others, which is not surprising in view of the two mechanisms. Figure 9.2 shows that even in the case of a constant fluence modification factor one may fail to observe the photoreversal at high UV fluences if mutation frequencies reach a plateau level. Because UV mutagenesis curves sometimes go through a maximum (see Section 9.5), the photoreversal of mutagenic UV lesions may even *increase* the observed mutation frequency at high UV fluence.

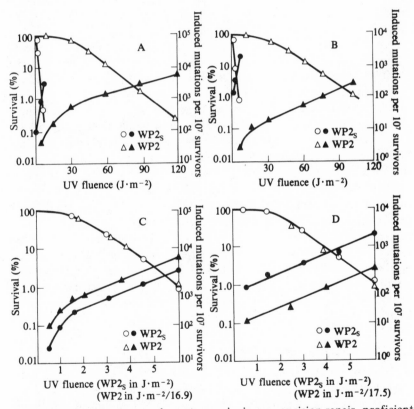

Figure 9.3. Inactivation and mutagenesis in an excision-repair proficient (▲, △) and an excision-repair deficient strain (●, ○) of *E. coli*. *Panel A*: Mutations to tryptophan independence. *Panel B*: Mutations to streptomycin resistance. *Panels C and D* show the same data as panels A and B, respectively, but the fluence scales for the two strains are altered so that their inactivation curves coincide (see text). (From E. M. Witkin, *Science 152*, 1345, 1966.)

9.3.2 Excision repair of mutagenic photoproducts

The previous conclusion that a considerable fraction of mutagenic UV photoproducts are pyrimidine dimers is consistent with extensive excision repair of mutagenic photoproducts. This was first brought out in 1965 by R. Hill, who found that the mutation frequency in an excision-repair deficient *E. coli* strain increases with UV fluence much more rapidly than in the corresponding repair-proficient strain. She also pointed out that at any given *survival* level the mutation frequencies in the two strains are similar. More detailed results, subsequently obtained by Witkin, are shown in Figure 9.3A and B. If the fluence scale for the repair-proficient strain is condensed relative to that of the repair-deficient strain so that the two survival curves coincide on the graph (Figure 9.3C and D), the induced mutation frequencies at equal survival levels differ far less than at equal fluences, but are not identical. Mutations to streptomycin resistance (panel D) still occur at an eightfold higher frequency in the repair-deficient strain, whereas mutations to tryptophan independence (panel C) occur only half as frequently as in the repair-proficient strain. As one would expect, considerably enhanced UV mutagenesis is also observed in excision-proficient cells when the repair processes are inhibited by caffeine (Figure 9.4).

Because excision repair reduces so drastically the frequency of induced mutations, its mechanism itself cannot be error-prone. It just abolishes the great majority of potentially mutagenic photoproducts so that only those left in the DNA will be available for error-prone repair processes. In view of the suggested mutagenicity of some nondimer photoproducts, it is not surprising that the mutation frequency still depends on whether the lesions available for error-prone repair are those left over by excision repair or are the total number of lesions formed by a much smaller UV fluence.

In agreement with the bacterial results, UV mutagenesis in phage T4 is also higher per unit fluence in excision-repairless mutants (v^-) than in the wildtype, as reported by J. W. Drake in 1973. Fluence modification factors due to v-gene repair (Section 7.5) are in this case identical for lethal and mutagenic lesions; therefore, at any given *survival* level the mutation frequencies are identical for wildtype and v^- phage.

9.3.3 Mutation frequency decline

If UV-irradiated auxotrophic cells of *Salmonella typhimurium* or *E. coli* are incubated for an interim period in a medium lacking amino acids (or containing chloramphenicol) before they are plated, the yield of prototrophy mutations is much reduced, whereas the cell survival remains unaffected. This phenomenon, first described by Witkin in 1956, is called *mutation-frequency decline*. It is not a general characteristic of UV mutagenesis, but limited to pleiotropic suppressor mutations, which occur relatively frequently. Among

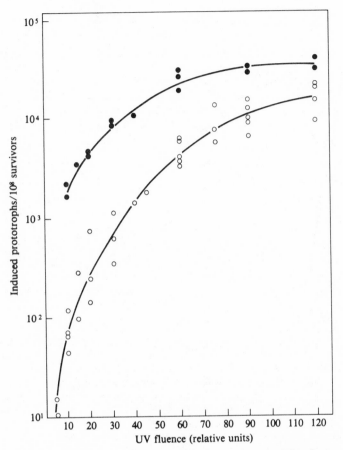

Figure 9.4. Effect of caffeine on UV mutagenesis. The frequency of UV-induced prototrophs within a culture of tryptophan-requiring *E. coli* cells is much increased by plating the cells on agar containing 0.1% caffeine (upper curve), compared with plating in the absence of caffeine (lower curve). (From E. M. Witkin, *Proc. 10th Int. Congr. Genetics*, Montreal, Canada, Vol. 1, 280, 1958.)

three possible explanations suggested by Witkin, the hypothesis of unstable premutational UV lesions best fits the current picture. Evidently, the premutational UV lesions are mostly pyrimidine dimers, and their instability is due to their potential to be excision-repaired. Mutation frequency decline due to incubation in media lacking amino acids is not observed in cell cultures previously grown in minimal media (supplemented with only the required nutrients), but only in those grown in a rich nutrient medium. Apparently lack of protein synthesis is the critical factor, which results either from the sudden amino acid deprivation or from the presence of chloramphenicol.

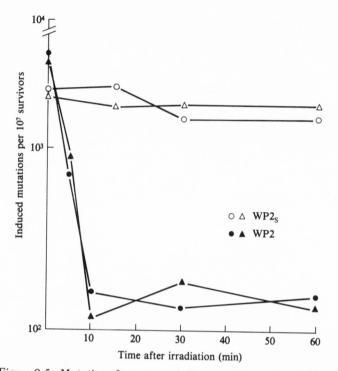

Figure 9.5. Mutation frequency decline in UV-irradiated *E. coli* cells. The percentage of induced mutations drops rapidly within the first 10 min after irradiation in wildtype cells (lower curve), but remains constant in excision-repair deficient cells (upper curve). (From E. M. Witkin, *Science 152*, 1345, 1966.)

The characteristics of the premutational state have been extensively studied in the laboratories of Witkin and C. O. Doudney. Of particular interest are experiments in which irradiated cells were first kept in low-mutation-yield medium and, at various later times, transferred to high-mutation-yield medium (usually Nutrient Agar plates). These conditions provide the best evidence that mutation frequency decline involves excision repair of mutagenic photoproducts. Figure 9.5 shows that the mutation frequency in irradiated wildtype cells (*E. coli* WP2) rapidly drops to less than $\frac{1}{10}$ as a result of holding them for only 10 minutes in liquid minimal medium prior to plating on solid nutrient-rich medium. No such effect is observed for the same type of mutations in irradiated cells of the excision-repair deficient derivative strain WP2$_s$. Furthermore, the data in Figure 9.6 indicate that mutation frequency decline in wildtype cells can be inhibited by the addition of acriflavine, a commonly used excision repair inhibitor. But addition of a large quantity of sodium deoxynucleate to the samples, which presumably binds most of the acriflavine, momentarily triggers the rapid decline.

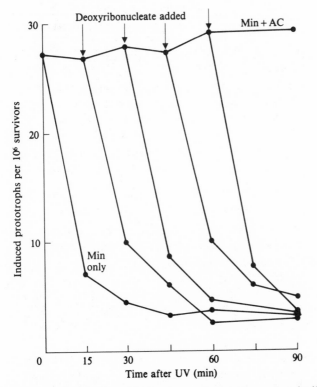

Figure 9.6. Effect of acriflavine on mutation frequency decline in *E. coli*. In the absence of acriflavine, mutation frequency decline follows the curve labeled "Min only." Permant presence of acriflavine in the minimal medium completely prevents mutation frequency decline ("Min + AC"), but the addition of sodium deoxyribonucleate at different times (arrows) results in immediate mutation frequency decline. (From E. M. Witkin, *J. Cell. Comp. Physiol. 58*, Suppl. *1*, 135, 1961.)

Mutation frequency decline is similar to liquid-holding recovery in that both are absent in excision-repair deficient cells and both are inhibitable in wildtype cells by caffeine, acriflavine, and crystal violet. Both phenomena represent a *reduction* of the UV response, mutagenic or lethal. Mutation frequency decline is also inhibited by other basic dyes with affinity for DNA, such as methyl green, methylene blue, and toluidine blue, but not by pyronine, which reacts selectively with RNA. On the basis of these considerations, one might wonder why a mutant not capable of mutation-frequency decline (*mfd⁻*) is as UV-resistant as the *mfd⁺* parent, indicating that it is excision-repair proficient. However, we should remember that excision-repair proficient strains also need not show liquid-holding recovery. Both phenomena, mutation frequency decline and liquid-holdings recovery, are the result of only *minor* modifications in the extent of excision repair. This is easy to realize if one compares, at a given UV fluence, mutagenesis in the excision-proficient *mfd⁻* strain with mutagenesis in an excision-repair deficient strain.

Mutation frequencies are also affected by postirradiation treatments other than those mentioned before; for example, various kinds of temperature changes. Like mutation frequency decline, such effects emphasize the existence of a *premutational state*, which is terminated at the time of *mutation fixation*. Time dependencies of various postirradiation treatments suggest that mutation fixation coincides with the first regular DNA replication after irradiation; only afterward is the final mutagenic response completely determined.

9.4 Mechanism of UV mutagenesis

Mutations are permanent alterations of the genetic information in DNA, determined by the sequence of nucleotide bases. Of various types of mutations, UV radiation causes predominantly *point mutations*, and less frequently specific kinds of *chromosomal aberrations* (Section 9.8). Point mutations typically consist of base pair substitutions, which result in either *missense* (i.e., an altered polypeptide chain) or *non-sense* (a truncated polypeptide chain). Alternatively, a single base pair can be added or deleted (*frameshift mutations*).

Obviously, such molecular changes cannot be immediate photochemical effects of UV radiation. They occur by secondary processes that the original photoproducts, predominantly pyrimidine dimers, initiate. As we have seen, a reduction in their number by photoenzymatic monomerization, or by their excision from the DNA molecule, lowers the frequency of induced mutation, whereas inhibition of such repair processes enhances it. As pyrimidine dimers remaining in DNA are lethal, one comes to the inevitable conclusion that mutations result from dimers that are neither monomerized nor excision-repaired, but that must have been otherwise eliminated by an *error-prone* process. This hypothesis, first publicized by Witkin, is now generally accepted, although more than one mechanism might exist, and details remain to be clarified.

Lack of UV mutagenesis in *recA⁻*, *exrA⁻*, or *lex⁻* strains of *E. coli* makes it likely that the error-prone process occurs in the course of recombination repair (postreplication repair), which ordinarily acts upon those dimers left in DNA after excision and/or photoenzymatic repair. It is now apparent that this repair mechanism consists of several possible pathways, of which the *recA* gene has the overall control. Mutations in genes affecting one or another of these pathways may, or may not, affect mutagenesis, but it is evident that the *exrA* gene (or *lexA* in *E. coli* K12) plays a central role in the mutagenic pathway(s). Strains carrying a mutation in this gene show no mutagenic response to UV radiation.

Our use of the term error-prone repair will not imply that an incorrect restitution of the DNA information by such a process is inevitable or at least common. The following rough calculation suggests that the probability per repair event for making an error may still be low. Exposure of *E. coli uvr⁻* cells to $1 \text{ J} \cdot \text{m}^{-2}$ of 254-um radiation produces an average of 60 dimers per chromosome. Because the chromosome consists of roughly 5×10^3 genes, any particular gene has a probability of the order of 10^{-2} of containing a dimer. If most *recA⁺* controlled repair events involved errors, and if at least 10 percent of these resulted in a phenotypically recognizable mutation, one would expect "forward" mutation frequencies of the order of 10^{-3} per average gene. Because mutation frequencies at this fluence are typically one to two orders of magnitude lower, the error rate per repair event cannot be high.

Unless the error-prone pathway amounts to only a very small fraction of the total *recA⁺* controlled repair, this calculation speaks against the simple idea that nucleotides are inserted at random where the dimer-containing template fails to provide the required information. More likely, a polymerase responsible for gap filling (whatever its mechanism) makes occasional but not frequent mistakes. Perhaps the recombination process itself, which of necessity involves some polymerization, is the mutagenic step. In the case of bacterial transformation, or in the case of fungi, genetic recombination indeed enhances mutability. Thus, any forced recombination steps associated with *recA⁺* controlled repair, which are prerequisite for survival of an individual, might enhance the frequency of mutants.

In recent years, a growing number of scientists have claimed that a repair process involving the *exrA⁺* function is *inducible* by the UV radiation itself. In agreement with such ideas, Sedgwick showed in 1975 that UV-irradiated *E. coli* cells synthesize a protein that is not observed in either unirradiated cells of the same strain, or in irradiated or unirradiated cells of *recA⁻* or *exrA⁻* cells. Such an inducible repair pathway (sometimes called *SOS repair*) might likewise be responsible for Weigle recovery of phage lambda and others (Section 7.9), which require intactness of the bacterial *recA⁺* and *exrA⁺* functions. We will here remember that under conditions of Weigle recovery the phage also shows enhanced UV mutagenesis, compared with infection of unirradiated host cells.

It is still somewhat early to integrate these observations into a general scheme. Possibly an operon becomes derepressed as a result of delayed protein synthesis in irradiated cells, which is likely in the case of prophage induction in UV-irradiated lysogenic bacteria (Section 10.4). More detailed knowledge in this regard will certainly contribute to our understanding of UV-induced mutagenesis; meanwhile, the overall picture can be discussed regardless of the inducibility or constitutivity of the proteins involved in error-prone repair.

The general hypothesis, namely, that recombination-related processes involved in postreplication repair are the essential cause of UV mutagenesis in bacteria, is supported by results obtained with bacteriophage. UV mutagenesis in T2 and T4 is observed specifically under conditions of multiplicity reactivation, for example, when intracellular phage are irradiated in the replicating state, or when extracellularly irradiated phages infect at high multiplicity. x^- mutants, which resemble bacterial *rec*$^-$ mutants with regard to their reduced UV repair and recombination frequency, produce at a given UV fluence only one-fourth of the mutations observed in a corresponding x^+ strain. Because of the greater UV sensitivity of x^- phages, the UV-induced mutation frequency would be even lower if compared with x^+ phages at the same survival level.

9.5 Kinetics of UV mutagenesis

Contrary to earlier hopes, kinetic data on UV mutagenesis have so far contributed little toward an understanding of the underlying processes. Nevertheless, they will gain importance as a criterion for their compatibility with ideas concerning the mechanism of UV mutagenesis. Because pyrimidine dimers are the predominant mutagenic photoproducts, whose formation follows single-hit kinetics, one would naively expect that the frequency of induced mutations parallels this kind of kinetics, provided that each dimer carries a *fixed* probability for causing a mutation. However, in most cases one finds different kinetics.

In studies on UV mutagenesis one measures directly the frequency of individuals affected, in contrast to inactivation studies, where actually the *complementary* function (i.e., survival) is recorded. Thus a single-hit mutagenesis curve would be expressed by

$$M_F - M_0 = 1 - e^{-\mu F} \tag{9.1}$$

where M_F is the relative mutation frequency obtained after irradiation with fluence F, M_0 is the relative frequency of spontaneous mutations, and μ is a rate constant expressing the mutagenic effectiveness of the radiation under given conditions. Because relative mutation frequencies are usually several orders of magnitude below unity, $e^{-\mu F}$ can be approximated by $1 - \mu F$, and induced mutagenesis would be expected to be a linear function with fluence:

$$M_F - M_0 \approx \mu F \tag{9.1a}$$

In the experimental reality, however, a plot of $M_F - M_0$ versus F usually increases more than proportionally with fluence, sometimes resembling a square function of the fluence:

$$M_F - M_0 = (\mu F)^2 \tag{9.2}$$

This function can be regarded as an approximation for either the two-target function $(1 - e^{-\mu F})^2$ or for the two-hit function $1 - e^{-\mu F} - \mu F e^{-\mu F}$ at low mutagenesis levels. To account for such a function on the basis of *induced* error-prone repair, it has been suggested that the first hit induces (or activates) the error-prone system, which, by repair of the second hit, actually causes the mutation. This explanation seems rather crude, but at present it can neither be ruled out nor supported.

The experimental data shown in Figure 9.3C and D are, at least at higher fluence, not compatible with two-hit or two-target kinetics. Because the ordinate scale is logarithmic, the straight parts of the curves express an exponential function of the type $M_F - M_0 = Ce^{F/F_m}$, where F_m is the fluence required for an e-fold increase of the mutation frequency, and C is a proportionality factor. In the same log plot, function (9.2) would be upwardly convex. But in any case, the considerably faster than fluence-proportional increase of the induced mutation frequency cannot be directly reconciled with the number of dimers originally formed. However, this quantity may be less critical for the result than the number of dimers undergoing error-prone repair, which unfortunately is not known. From the data it is only safe to say that the probability for making an error in the course of repair of a lesion increases with the number of lesions originally produced.

If one wishes to compare UV-induced mutation frequencies in cells exhibiting different repair capacities, one can plot the mutation frequency either as a function of UV fluence or as a function of survival. Preference for one kind of plot or for the other depends on the information the data are supposed to provide. For example, it is clear from the presentation of data in Figure 9.3A and B that at a given fluence (i.e., at a given number of original photoproducts) UV mutagenesis is much greater in excision-repair deficient cells than in excision-repair proficient cells, suggesting that excision repair removes potentially mutagenic photoproducts in an error-free manner. Presentation of the same data in the form of Figure 9.3 C and D tells us that identical numbers of UV photoproducts resulting either from a small UV fluence without excision repair, or from a much greater fluence with subsequent extensive excision repair, are qualitatively different because they lead to different mutation responses. Unless we recognize which type of plot is *suitable* for answering a particular question, misinterpretations may be the consequence. Sometimes it is advantageous to plot data in both fashions because the conclusions may complement each other. On the other hand, contradictions between conclusions drawn from the two kinds of plots serve as a warning, indicating that one of them must be wrong. Quantitative logic can usually uncover the reason for misinterpretation.

A kinetic paradox, sometimes observed at high UV fluences, is the leveling-off or passing through a maximum of the mutation frequency, in spite of continuing dimer formation. A possible reason for this is *population heterogeneity* with respect to repair, for which there is evidence in all those cases where the survival curves end in a more resistant "tail" at high fluences. Apparently a small fraction of the population is capable of repairing its lethal photoproducts more extensively than the remainder. If this were the case for *error-free* repair of mutagentic photoproducts, the following example illustrates how it could explain a maximum of mutation induction.

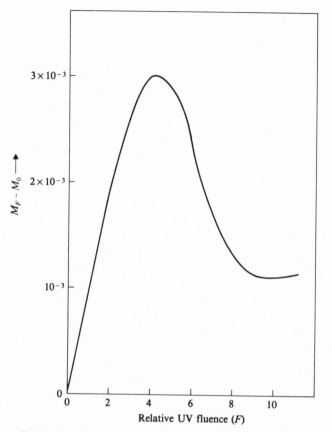

Figure 9.7. Expected function for UV-induced mutagenesis according to equation (9.3.)

We assume, for simplicity, that a cell population S_0 consists of two sub-populations, S_{01} and S_{02}, both of which are inactivated and mutate according to one-hit kinetics. If the relative sizes of the subpopulations are $S_{01}/S_0 = 0.99$ and $S_{02}/S_0 = 0.01$, and the UV inactivation rate constants are c_1 and $c_2 = 0.1c_1$, respectively, the survival of the whole population is expected to be

$$S/S_0 = 0.99\, e^{-c_1 F} + 0.01\, e^{-0.1c_1 F}$$

If the two subpopulations likewise differ 10-fold in their mutation rate constants μ, with μ_2 being $0.1\mu_1$, the frequency of mutants among the survivors would be

$$M_F - M_0 \approx \frac{\mu_1 F \cdot 0.99 e^{-c_1 F} + 0.1\mu_1 F \cdot 0.01 e^{-0.1 c_1 F}}{0.99 e^{-c_1 F} + 0.01 e^{-0.1 c_1 F}} \tag{9.3}$$

Such a function is plotted in Figure 9.7 for values of $c_1 = 1.0$ and $\mu_1 = 10^{-3}$. In qualitative terms, the maximum is due to the fact that at low fluences the mutants represent essentially the more mutable and more UV-sensitive majority of the cells, whereas at higher fluence the survivors come essentially from the small, UV-resistant fraction of cells showing low mutability. Obviously, *within* each subpopulation the mutation frequency would steadily increase with fluence, though at 10-fold different rates.

9.6 Delayed appearance of UV-induced mutations

Experimental work by M. Demerec and collaborators in the 1940s showed that UV-induced mutations are often not detectable until several cell generations after irradiation. Such delayed effects were observed in *E. coli* for many different types of mutations, notably resistance to phage or antibiotics, auxotrophy, and reversions to prototrophy, among others. In an extreme case, up to 13 cell generations were required for the mutation rate to return to the spontaneous level. Later studies in various laboratories revealed several potential causes for the delayed appearance of mutations:

(a) The molecular alteration in DNA, which constitutes the mutation, is delayed (*mutational delay*). The suggested mechanism of UV mutagenesis explains mutational delay simply by the indetermination existing prior to mutation fixation as a result of error-prone repair processes.

(b) The phenotype is expressed considerably later than the occurrence of the mutational event (*phenotypic delay*). Altered biosynthetic characteristics of the mutant genotype usually require only a short time to establish a new phenotype. But bacterial resistance to phage T1, where the delayed appearance of induced mutations was first noticed, is due to the *absence* of phage receptors in the bacterial cell wall. Thus, expression of the new phenotype requires that the existing receptors be diluted out or otherwise lose their specificity in subsequent cell divisions.

(c) If the induced mutation is recessive, and cells contain more than one set of genetic information, segregation of the mutated genome leading to homozygosity is a requirement for phenotypic expression (*segregational delay*). *E. coli* mutations leading to inability of lactose fermentation can be recognized on eosine–methylene blue–lactose agar, where mutants form light-colored colonies in contrast to dark-red wildtype colonies. UV-induced mutations frequently give rise to sectored colonies (Figure 9.8), indicating that only a fraction of the descendants from the original irradiated individual have segregated to display the mutated phenotype. There is a strong correlation between the ratio of intact/sectored lactose-negative colonies and the ratio of uninucleate/multinucleate cells in the UV-irradiated population, which is additional evidence for the postulated segregational delay.

(d) Phenotypic expression of an induced mutation requires at least DNA

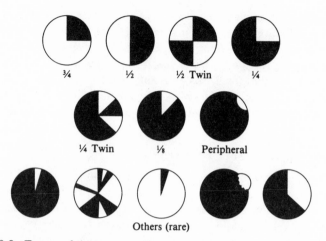

Figure 9.8. Types of lactose-negative sectored colonies obtained after UV irradiation and plating on eosine-methylene blue agar. (From E. M. Witkin, *Cold Spring Harbor Symp. Quant. Biol. 16*, 357, 1951.)

replication, and often cell division(s). Therefore, the delayed appearance of mutations may be aggravated by variations in the delay of these functions after UV irradiation (*irregularity delay*).

Any one of these factors, and sometimes several combined, have been shown to contribute to the delayed expression of UV mutagenesis. For determination of the UV-induced mutation frequency it is mandatory to know the conditions and the time dependence of their phenotypic expression. In using UV radiation as a convenient mutagen for the creation of certain desirable genotypes, it is advantageous to allow the majority of survivors to go through several cell divisions in order to express their UV-induced mutations.

9.7 Indirect UV mutagenesis

UV mutagenesis usually arises from photoproducts formed by *direct* interaction of photons with the DNA. However, near-UV radiation particularly can also act mutagenically in an *indirect* way, that is, by formation of photoproducts in other cell components or in the medium, which then serve as chemical mutagens. For example, W. S. Stone and collaborators (1947) increased mutation frequencies in unirradiated *Staphylococcus aureus* cells by inoculation into *UV-irradiated* nutrient broth media, but not by inoculation into irradiated glucose-salts media. Similar effects were found after pretreatment of broth with hydrogen peroxide, which failed to occur in the presence of catalase, suggesting that the observed indirect UV effects involve organic peroxides.

The effects of both UV-irradiated and peroxide-treated media resemble those of direct UV interaction in the temperature dependence of the inactivation effect in *E. coli* B cells (Section 8.3), and in the greater resistance of strain B/r. More recent work by Ananthaswamy and Eisenstark (1976) provided evidence that the inactivating and mutagenic effects of near-UV radiation (e.g., 365 nm) are due to tryptophan photoproducts, among which hydrogen peroxide plays the predominant role. Hydrogen peroxide as well as the tryptophan photoproducts inactivate particularly *recA⁻* strains of *E. coli*, probably by interfering with the resynthesis step in excision repair. Because bacterial mutagenesis by near-UV radiation (365 nm) is much more frequent than would be expected from the (very low) DNA absorbance, it is likely that these mutagenic effects also involve indirect mechanisms.

A possibly indirect mutagenic effect by *far*-UV radiation was reported by Kada and Marcovich in 1963. *Hfr* chromosomes of UV-irradiated *E. coli* K12 cells undergo mutation with increasing time elapsed before their conjugational transfer to F⁻ cells. Conversely, the mutation frequency in *Hfr* chromosomes transferred from unirradiated cells into preirradiated F⁻ cells increases with the preirradiation UV fluence. Because these effects are not observed with X-rays, the results may represent an indirect UV action through the formation of cytoplasmic photoproducts. This seems more likely than the possible explanation of the effect by induced, error-prone repair processes because the (unirradiated) *Hfr* chromosome entering the irradiated F⁻ cell should not require such repair.

In general, mutations caused by 254-nm radiation are due to pyrimidine dimer formation by direct photoabsorption in DNA. However, the possible role of an indirect effect must be considered when dimer repair is unusally extensive, or when a lack (or low degree) of photoenzymatic repair indicates the predominance of other mutagenic UV photoproducts.

9.8 UV induction of chromosomal aberrations

Compared with ionizing radiations, ultraviolet light is rather inefficient in producing chromosomal aberrations in eukaryotic cells. H. J. Muller and co-workers showed that a UV fluence resulting in several percent of sex-linked lethal mutations in *Drosophila* produces no *translocations*. For comparison, an X-ray dose yielding a similar percentage of sex-linked lethals produced at least 65 detectable translocations. Even *minute rearrangements*, which are scored with a special technique at high frequency after X-ray treatment, are not observed after UV irradiation. The only type of chromosomal aberrations induced at a substantial rate in *Drosophila* seem to be minute interstitial *deletions*.

The primary objects for investigations on UV-induced chromosomal aberrations in higher plants have been *Tradescantia* and *Zea mays* (maize, or corn).

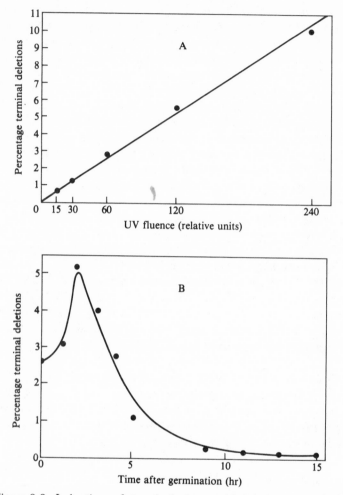

Figure 9.9. Induction of terminal chromatid deletions in *Tradescantia* by 254-nm UV radiation, using the pollen tube technique. *Panel A* shows the linear increase with fluence (60 sec corresponding to approximately 100 $J \cdot m^{-2}$). *Panel B* shows the dependence of the percentage of aberrations on the time after germination, at which the chromosomes were exposed to approximately 200 $J \cdot m^{-2}$. (From C. P. Swanson, 1942 and 1943; see Swanson and Stadler in: *Radiation Biology*, A. Hollaender, ed., Vol. II, McGraw-Hill, New York, 1955, pp. 249–84.)

In maize, aberrations can be detected as losses of linked genes determining endosperm characters, whereas in *Tradescantia* direct cytological observation of the irradiated chromosomes is made possible by the pollen-tube technique (Section 9.1). As shown in Figure 9.9A, terminal chromatid deletions are produced in *Tradescantia* in considerable number. The increase is a linear function of the UV fluence if the chromosomes are exposed at a fixed time after pollen germination, suggesting that the aberrations result from a single photochemical event. However, at a given fluence, the percentage of chromatid deletions depends greatly on the chromosomal stage; it declines to very low values with increasing condensation during prophase (Figure 9.9B). Presumably these structural changes affect the probability for chromosomal breakage as a result of photon absorption.

None of the terminal deficiencies includes both chromatids, and rarely do there occur in *Tradescantia* other chromosomal aberrations, such as translocations. In maize, the results obtained by various authors are similar; no translocations are found, and all the observed deletions are terminal. Only studies on ring chromosomes in maize pollen revealed frequent breakage after UV irradiation, and a high incidence of UV-induced translocations was found in *Gasteria*; but such observations are exceptions rather than the rule. In general, UV radiation induces at a considerable rate only a few specific types of chromosomal aberrations. The reason could be that ultraviolet light, in contrast to ionizing radiations, rarely causes DNA strand breakage (Section 3.3.5), and that the relatively frequent cases of chromatid breaks are the result of secondary processes in the course of repair.

It is interesting to notice that in *Tradescantia* a combined X-ray and UV treatment, regardless of the sequence, induces chromatid deletions at the same frequency as either one of the two treatments alone. The frequency of X-ray-induced isochromatid deletions and translocations is even *reduced* to one-third or less by additional UV radiation; the significance of this apparent antagonism is emphasized by similar effects observed in *Drosophila* and maize. As a possible explanation, very effective repair of UV damage could coincidentally abolish X-ray lesions for which they otherwise have no specificity.

10 UV effects of macromolecular syntheses, genetic recombination, lysogeny, and related phenomena

10.1 DNA synthesis

An immediate consequence of UV irradiation in cells is a reduced rate of synthesis of macromolecules. An early observation by Kelner (1953), later confirmed in many laboratories, was that the effect is much more drastic on DNA synthesis than on synthesis of either RNA or proteins. Figure 10.1 shows that a moderate fluence (20 J · m⁻² at 265 nm) brings DNA synthesis in *E. coli* B cells to an almost complete halt for a certain period, whereas RNA and protein syntheses continue at reduced rates.

Many investigations addressed to these matters provided evidence that the extent of DNA-synthesis inhibition depends in a complex manner on the UV fluence and on the repair capabilities of the cells investigated. For example, strain B_{s-1} (*uvrB exrA*), a repair-deficient derivative of *E. coli* B, shows complete cessation of DNA synthesis for at least 60 to 80 minutes after exposure to only 1 J · m⁻², whereas the very UV-resistant bacterium *Micrococcus radiodurans* shows a delay of only 20 to 40 minutes even after 100 J · m⁻². Studies by Doudney and Young (1962) showed with excision-repair proficient cells that the time period between irradiation and resumption of DNA synthesis increases with UV fluence and reaches a maximum; any further increase in fluence lowers, in addition, the synthesis rate after resumption.

For some time, it was thought that semiconservative DNA replication is completely blocked by a pyrimidine dimer in the template strand, and cannot continue unless the dimer is either photoenzymatically monomerized or excised, and an undamaged sequence is reconstituted by repair synthesis. However, it is now well established by studies on recombination repair (Section 7.6) that semiconservative DNA replication blocked by a dimer can resume at some point beyond the dimer, leaving a gap of the order of a thousand nucleotides in length in the newly synthesized strand. At a fluence producing more than a thousand dimers per *E. coli* chromosome, one would indeed expect relatively little DNA synthesis until excision repair had taken place. The implication that pyrimidine dimers are the primary cause of inhibited

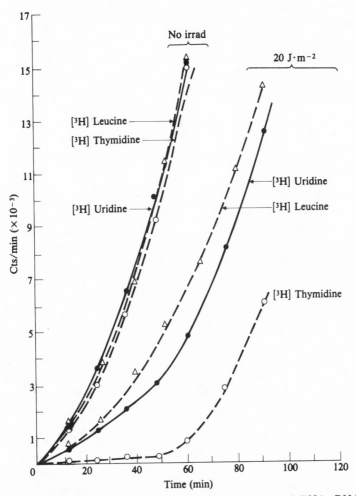

Figure 10.1. Effect of UV radiation on the synthesis of DNA, RNA, and protein in *E. coli*, measured by incorporation of ^{3}H-labeled precursors thymidine, uridine, and leucine, respectively. (From P. A. Swenson and R. B. Setlow, *J. Mol. Biol. 15*, 201, 1966.)

DNA synthesis is strengthened by the fact that photoreactivating treatment of irradiated cells shortens the cessation period and enhances the rate of DNA synthesis after resumption, if it was affected by the radiation.

10.2 RNA and protein synthesis

The smaller effects of UV radiation on RNA and protein synthesis, compared with DNA synthesis, can be equally well explained by the presence of pyrimidine dimers in DNA. Specifically, unless the total number of dimers in a DNA molecule is rather large, synthesis of messenger RNA, ribosomal RNA, or transfer RNA in a cell should be affected only to the extent that the corresponding cistrons (or group of cistrons) carry at least one dimer. In consequence, protein synthesis should be little affected. Inhibition of both RNA and protein syntheses becomes gradually more noticeable with increasing UV fluence, as the average "concentration" of photoproducts in DNA rises.

Pyrimidine dimers in DNA apparently block the transcription by DNA-dependent RNA polymerase as much as they block replication by DNA polymerase. The consequence is *shortened* messenger RNA molecules, containing incomplete information. According to Bremer and coworkers a fluence of $100 \text{ J} \cdot \text{m}^{-2}$ of 254-nm radiation produces in *E. coli* DNA an average of one transcription-terminating lesion per thousand base pairs, which is roughly the number of pyrimidine dimers produced in one DNA strand of this length. This result strongly suggests that any dimer located in the transcribed DNA strand terminates RNA chain elongation. The decreasing RNA synthesis rate observed with increasing UV fluence suggests that RNA polymerase molecules can be either delayed in their release from UV-irradiated DNA or irreversibly bound.

UV effects on protein synthesis in *E. coli* complement the picture on RNA synthesis. As one would expect from the reduced rate of RNA-chain initiation in irradiated cells, the rate of polypeptide chain production is similarly affected. In addition to polypeptide chains of normal length, truncated chains are formed in the translation process, most likely as a consequence of incomplete information provided by the shortened m-RNA molecules.

Termination of transcription by a pyrimidine dimer in DNA can explain the UV inactivation of specific gene functions, which has been extensively studied in bacteriophage. Such experiments are basically carried out as follows. Cells are multiple-infected with an *unirradiated* phage mutant lacking a vital function [for example, T4*r*II mutants infecting K12(λ) cells, or *am* mutants infecting strains lacking an amber suppressor], and single-infected with a homologous *UV-irradiated* phage providing the vital function. Thus the mixedly infected cells will produce progeny as long as the critical gene function has not been destroyed by the irradiation. Consequently, loss of plaque formation by these complexes measures directly the UV sensitivity of particular gene functions, as was first shown experimentally in 1959 by D. R. Krieg.

Although some gene functions are quite UV-resistant (for example, the v^+ gene function of T4; see Section 7.5), usually their UV sensitivity is greater than expected from the size of the gene, relative to that of the total genome. The reason is that not only UV lesions within the particular gene, but also in its neighborhood, can interfere with transcription and thereby inactivate the gene function. This is the case if several genes are transcribed sequentially as a unit, and a pyrimidine dimer terminates transcription in a DNA region proximal to the point of messenger-RNA initiation. Therefore, the closer a gene's location is to the initiation point, the less UV-sensitive should its function be. Theoretically, if several genes A, B, C, D ... are transcribed in this sequence, the sensitivity of gene function B is expected to approximate the relative size of genes (A + B), and the sensitivity of the gene function C should correspond to the relative size of genes (A + B + C), and so forth. This principle has been successfully applied since 1970 by W. Sauerbier and collaborators for determination of the sequence of genes, their grouping in transcription units, and the direction of the transcription process for various phages.

The delay in cellular synthesis of macromolecules after UV irradiation can have various nonlethal, biologically detectable consequences. Some of the well investigated ones will be discussed in the following sections.

10.3 Delay of growth and cell division

10.3.1 Prolongation of the latent period in phage multiplication

The multiplication cycle of a bacteriophage is well characterized by a *one-step growth curve*. This is a plot of the number of plaque-forming units (i.e., the sum of infected, phage-producing cells *and* extracellular viable phage) as a function of time after infection, which one can obtain in an experiment originally designed by Ellis and Delbrück in 1939. As illustrated in Figure 10.2, after a certain time period (called the *minimum latent period*) the number of plaque-forming units increases sharply (*rise period*) and then levels off again. Evidently the minimum latent period defines the time required until the first cells burst and release phages. The steepness of the increase during the rise period reflects the degree of synchrony in phage multiplication and lysis of the infected cells, and the factor of increase corresponds to the *average burst size*.

In 1944 Luria observed that UV irradiation of extracellular phage lengthens both the minimum latent period and the rise period, the latter indicating greater individual variations in the prolongation of the latent period. For irradiated phages T1 and T7 the minimum latent period increases roughly proportionally to UV fluence and is doubled at a survival of about 10^{-3}. The fact that the phage particles emerging from such delayed cell lysis behave in the next growth cycle like unirradiated phage, indicates that the prolongation of the latent period is a *temporary, nonhereditary* UV effect. It is a fairly generally observed consequence of phage irradiation, varying in its extent with the type of phage.

The action spectrum for the delayed lysis of phage-infected cells suggests nucleic acid absorption as the cause, and the reduction of the effect as a result of photoreactivating treatment indicates that pyrimidine dimers are the relevant photoproducts. One might, therefore, suspect that the prolonged

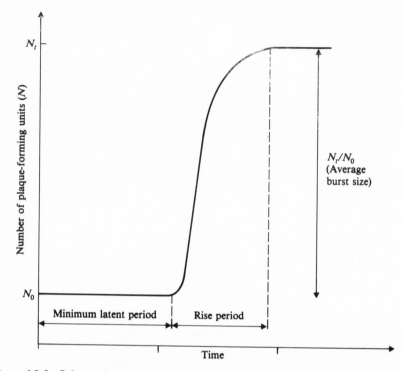

Figure 10.2. Schematic representation of a one-step growth curve of a T-even bacteriophage (see text).

latent period represents the time requirement for the excision repair of dimers, but this is not the case, as can be concluded from the following results.

Figure 10.3 shows that UV-irradiated phage T1 infecting excision-repair deficient cells show a much more dramatic prolongation of the latent period than the same phage infecting repair-proficient cells. For an equal extent of prolongation, phages infecting repair-proficient cells must receive roughly four to five times the fluence required for phage infecting repair-deficient cells. This suggests that *unrepaired* pyrimidine dimers, rather than the excision repair itself, cause the delay in intracellular phage propagation. In the T-even group, T1, and lambda phages, a minor fraction of UV lesions are known to be repaired by recombination-related processes; it is not unlikely that their additional time requirement results in an extended latent period.

10.3.2 Delay of bacterial cell division

Resting bacteria transferred into new growth medium always show a lag before they start dividing; this period is considerably longer than the time interval between two subsequent cell divisions during exponential multiplica-

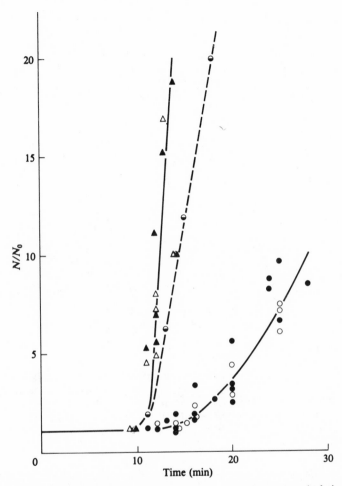

Figure 10.3. UV-induced prolongation of the latent period in phage T1 infecting either excision-repair proficient (open symbols) or deficient (closed symbols) host cells. Compared with unirradiated phages (steep solid curve), approximately equal prolongations are found after 23 J · m⁻² and infection of repair-deficient cells and after 100 J · m⁻² and infection of repair-proficient cells, whereas irradiation with 23 J · m⁻² and infection of repair-proficient cells results in little prolongation (dashed curve). (From W. Harm, *Photochem. Photobiol. 4*, 575, 1965.)

tion. In UV-irradiated cells this lag period is lengthened, as was first observed in 1938 by Hollaender and Duggar and later confirmed in many laboratories. A lag period after irradiation is also observed with exponentially growing cells, but even after the first postirradiation division, they multiply at the rate of unirradiated cells. Figure 10.4 shows that the bacterial lag period, like the latent period of phage, increases proportionally with UV fluence; it reaches

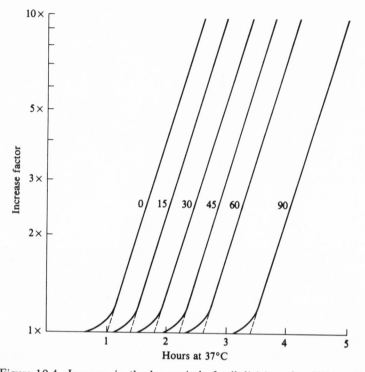

Figure 10.4. Increase in the lag period of cell division after UV irradiation of stationary phase *E. coli* cells. The numbers on the curves are seconds of UV irradiation at a fluence rate of 0.6 J · m^{-2} · sec^{-1}. The extent of displacement of the curves shows direct proportionality of the increase in lag period with fluence. (From E. M. Witkin, *Proc. 10th Int. Congr. Genetics*, Montreal, Canada, Vol. 1, 280, 1958.)

a maximum value at still higher fluence. Photoreversibility of this effect indicates that it is caused by pyrimidine dimers. Often an extended lag period of irradiated cells represents at the same time a growth delay; that is, the increase not only in the number of cells but also in total cell mass in the culture is slower than in a corresponding culture of unirradiated cells.

Again, the time requirement for excision repair is not responsible for the division delay because *uvr*⁻ cells show a very substantial extension of the lag period even at the rather low UV fluences applicable. It is more likely that the slowdown is caused by unexcised pyrimidine dimers, which interfere with DNA replication and which must be dealt with by recombination repair or a related process. This interpretation is compatible with the observation of no division delay in *uvrA*⁻ *recA*⁻ *E. coli* cells, where solely dimer-free cells survive. It should be admitted, however, that in this case the UV fluence applicable is extremely low, and a test for division delay with a *uvr*⁺ *recA*⁻ strain, if negative, would provide better evidence.

10.3.3 Filament formation

An extreme case of division delay in UV-irradiated cells, described by Witkin in 1947 for *E. coli* B, is the formation of filaments, or "snakes." Primarily at low fluences, where the majority of cells survive, they undergo a tremendous elongation, reaching many times the length of normal cells, before a certain percentage of them finally divide and form a macroscopically visible colony. Microphotographs of such filaments are seen in Figure 10.5; their mass and the contents of DNA, RNA, and proteins can be equivalent to many cells.

Filament formation represents grossly delayed cell division *without* correspondingly delayed growth of cell mass. This is probably due to a regulatory malfunction for cross-wall formation of the cells, whose details are not yet understood at the molecular level. Filaments are characteristic of *E. coli* strains carrying a certain mutation (called *fil* in strain B, or *lon* in K12); they are much less frequent in the B-derivative strain B/r, which possesses a *fil*

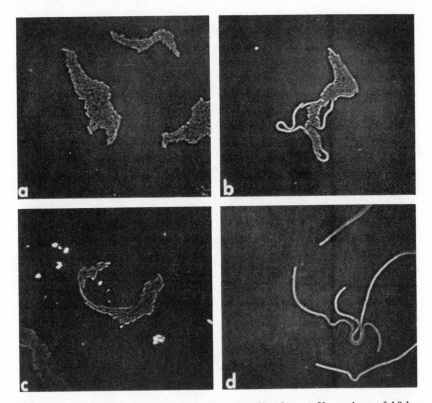

Figure 10.5. Filament formation in *E. coli* cells after an X-ray dose of 10 kr, resembling the effect obtained after UV irradiation. (a) Unirradiated wildtype cells; (b) unirradiated cells of a filament-forming (*lon*) mutant strain; (c) irradiated wildtype cells; (d) irradiated *lon* cells. (From H. I. Adler and A. A. Hardigree, *J. Bacteriol. 87*, 720, 1964.)

suppressor, and are absent or rare in most wildtype strains. H. I. Adler and co-workers (1966) showed that extracts from nonfilamentous cells contain a division-promoting principle, which upon addition to filaments induces cell division and leads to enhanced survival.

The photoproducts responsible for filament formation are pyrimidine dimers in DNA, which is indicated by the DNA-like action spectrum and by the reduced filament formation resulting from photoreactivating treatment. Filaments are absent after irradiation at high UV fluence, where the survival is below 10^{-3}. This is not surprising because in this case the macromolecular syntheses and the resulting growth of cell mass are presumably as severely affected as the capability of cross-wall formation.

The UV survival of cells of a filament-forming strain depends greatly on postirradiation conditions. Any postirradiation treatment reducing the tendency of filament formation likewise enhances the chance for cell survival. Examples are the presence of pantoyl lactone or chloramphenicol, incubation at 44–45°C, or plating on glucose-salts minimal media. Filament formation of *E. coli* B is strongly correlated with the greater sensitivity of this strain compared with B/r or other wildtype strains. This holds not only for UV radiation, but is characteristic of many other DNA-damaging agents, such as ionizing radiations, peroxides, nitrous acid, mitomycin C, nitrogen mustard, crystal violet, and others. All of these agents inactivate B cells more strongly than B/r cells, and lead to the formation of filaments in strain B but hardly in B/r. *E. coli* B cells in the filamentous state are themselves much more UV-sensitive than normal cells.

10.3.4 Growth delay

UV-irradiated cells in a bacterial culture often show a slower increase in their total mass (regardless of cell divisions) than unirradiated samples. This effect, called growth delay, is commonly observed after UV irradiation and also after many other kinds of adverse treatment. Of course, if the radiation inactivates a substantial fraction of the population, a growth delay of the culture would be trivial if it were solely due to failure of a fraction of the population to increase in mass and to divide. However, this is not the case. Growth delay is as well expressed by noninactivated cells, which is particularly evident at low UV fluences. The wavelength dependence of growth delay in the far-UV region indicates that nucleic acid absorption is the primary cause, which comes as no surprise in view of the pyrimidine dimer effects on macromolecular syntheses (see Sections 10.1 and 10.2).

An interesting aspect of UV-induced growth delay in cells is the relatively high effectiveness of near-UV wavelengths, which is maximal at about 340 nm. Figure 10.6 shows that the near-UV action spectrum for growth delay coincides with those for photoprotection and indirect photoreactivation, suggesting that all three phenomena are based on similar photochemical

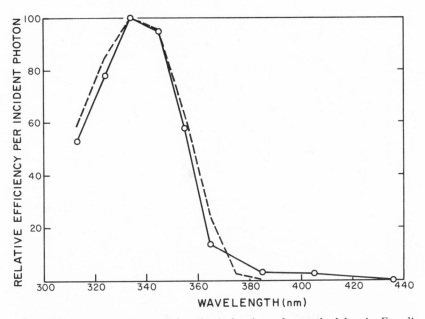

Figure 10.6. Action spectrum for the induction of growth delay in *E. coli* B by near-UV radiation (solid line). The action spectrum for photoprotection (dashed line) is shown for comparison. (Drawn from tabulated data of J. Jagger, W. C. Wise, and R. S. Stafford, *Photochem. Photobiol. 3*, 11, 1964.)

reactions. Recent studies by Ramabhadran and Jagger showed that a near-UV fluence causing extensive growth delay leads to a complete (though temporary) cessation of net RNA synthesis, whereas DNA and protein syntheses proceed at reduced rates. Because the action spectrum for growth delay coincides with the absorption spectrum of the base 4-thiouracil present in a number of transfer RNA species, they proposed the following hypothesis: A known photochemical reaction, namely, the formation of a photoadduct between 4-thiouridine in the 8-position of the RNA molecule and cytidine in the 13-position, leads to inactivation of transfer RNA. This in turn causes a complete shut-off of net RNA synthesis by the same regulatory mechanism by which it is shut off in the case of amino acid starvation, and thus causes the growth delay.

This hypothesis is strongly supported by the following evidence. It is known that only *stringent E. coli* strains (*rel$^+$*) discontinue their RNA synthesis in the case of amino acid starvation; *relaxed* strains (*rel$^-$*) do not. The experimental data presented in Figure 10.7 show that the same is true in the case of near-UV irradiation of cells: In *rel$^-$* strains the RNA synthesis is little affected and growth delay is almost absent. Furthermore, after amino acid starvation as well as after near-UV exposure of *rel$^+$* cells, resumption of RNA synthesis is stimulated by chloramphenicol.

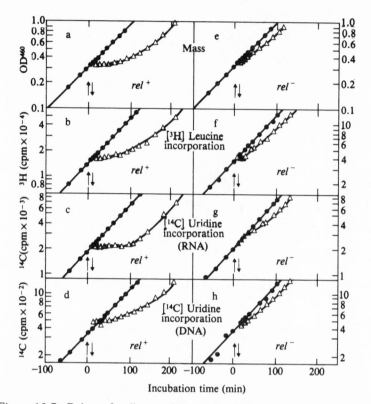

Figure 10.7. Delay of cell growth, protein, RNA, and DNA synthesis in *E. coli* cells after exposure to 2×10^4 J·m^{-2} broadband radiation (320–405 nm) from a blacklight lamp. As described in the text, the effects in *rel*$^+$ cells (a–d) are much different from those in *rel*$^-$ cells (e–h). (From T. V. Ramabhadran and J. Jagger, *Proc. Nat. Acad. Sci. U.S. 73*, 59, 1976.)

10.4 Lysogenic induction and related phenomena

Lysogenic bacterial strains, carrying phages in some latent form, were discovered shortly after the bacteriophages themselves, but their real nature remained obscure for more than three decennia. In 1950, Lwoff and collaborators showed in their Nobel prize-winning work that *lysogeny* is a hereditary trait of every cell in a lysogenic strain, namely, to carry a latent copy of phage (*prophage*) along with its other genetic determinants. With a low probability (usually of the order of 10^{-3} to 10^{-6} over the life span of a cell) the prophage converts to the vegetative state, in which it multiplies rapidly and produces progeny that will finally destroy the host cell. The remaining lysogenic cells in the culture are immune to the liberated phage particles, but the latter can infect, lytically or lysogenically, cells of other

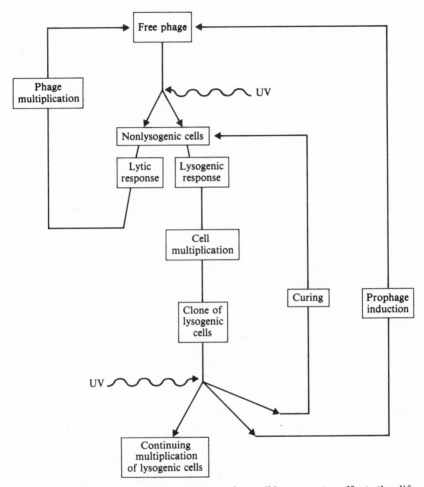

Figure 10.8. Schematic representation of possible ways to affect the life cycle of a temperate phage by UV radiation (see text.)

bacterial strains. The capability of establishing lysogeny is characteristic only of *temperate* phages, in contrast to *virulent* phages.

UV radiation has various possible effects on lysogeny, as illustrated schematically in Figure 10.8. (1) Irradiation of extracellular phage can alter the probability for lysogenization of a sensitive cell; (2) irradiation of lysogenic cells can greatly enhance the probability for transition from prophage to the vegetative state (lysogenic induction or prophage induction); and (3) irradiation of lysogenic cells can result in loss of the prophage without killing the cell (*curing*). Lysogenic inducibility, the most important of these effects, is due to a property of the prophage, and therefore occurs only in some bac-

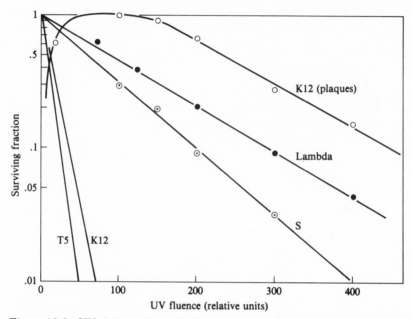

Figure 10.9. UV inactivation of *E. coli* cells lysogenic for phage lambda (K12) and of nonlysogenic cells (S). Curves representing the fraction of K12 cells liberating phages after UV induction (K12, plaques) as well as the survival curves of free phages lambda and T5 are shown for comparison. (From J. J. Weigle and M. Delbrück, *J. Bacteriol. 62*, 301, 1951.)

terial strains (inducible strains) but not others. As a rule, inducible lysogenic cells are more UV-sensitive than the corresponding nonlysogenic cells because of their *additional* chance of being killed by the induced prophage. An example is shown in Figure 10.9. The survival curve of strain K12(λ) is considerably steeper than that of K12S (which lacks the prophage lambda). The survival of extracellular lambda and the fraction of K12(λ) cells liberating viable phage upon UV induction are shown for comparison.

The marked difference between the K12(λ) and K12S survival is not surprising, as we see that a relatively low UV fluence induces nearly all of the K12(λ) cells. At higher fluences the fraction of cells liberating viable phages decreases at a rate resembling that for inactivation of extracellular phages. This may reflect inactivation of the prophage, although the degree of similarity may be fortuitous because the phage survival should be affected by Weigle recovery (Section 7.9) as well as by the loss of the cells' capacity to propagate phage lambda.

Photoenzymatic repair affects prophage induction in the expected manner; that is, an induction curve resembling that in Figure 10.9 is obtained at proportionally higher radiation exposure if it is followed by photoreactivating

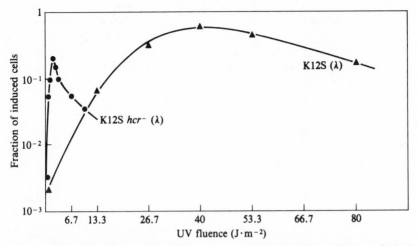

Figure 10.10. UV induction of λ-lysogenic cells of an excision-repair proficient [K12S(λ)] and an excision-repair deficient [K12S *hcr⁻*(λ)] strain of *E. coli.* (From W. Harm, *Photochem. Photobiol. 4*, 575, 1965.)

treatment. Excision repair has similar effects: The UV fluence required for maximum induction of lysogenic excision-repair deficient cells is much lower than required for the corresponding excision-repair proficient cells, as shown in Figure 10.10. Likewise, the fraction of phage-producing cells beyond the induction maximum decreases much more rapidly in a repair-deficient strain because the prophage is inactivated at a greater rate.

Thus, there is little doubt that the prophage-inducing photoproducts are unrepaired pyrimidine dimers in DNA, but the secondary processes leading to the induction are not known in detail. The prophage state is established and maintained by a sufficient amount of repressor, which prevents the phage from carrying out its vegetative functions. Repressor molecules for inducible phages, in contrast to those for noninducible phages, are unstable; therefore, any interference with their continuous production tends to derepress (induce) the prophage. The problem still remains of explaining how the implied repressor shortage can be caused by a relatively low UV fluence.

UV exposure of *extracellular* phages enhances their likelihood of lysing sensitive host cells rather than lysogenizing them. This effect is typical of inducible phages only, and a possible relationship with prophage induction is suggested by the fact that in both cases UV irradiation favors the vegetative state vis-à-vis the prophage state. Both lysogenization and prophage induction involve recombination between bacterial and phage genome, requiring the phage genes *int* and *xis*.

Exposure of lysogenic cells to rather high UV fluences can result in *curing* of some of the survivors, as was discovered in 1951 by E. Lederberg. She found that after heavy irradiation of *E. coli* K12 a few of the isolated colonies had become sensitive to a then unknown phage (henceforth called *lambda*) present in the culture. Cured cells may have lost their prophage spontaneously and may be selected by UV irradiation because of their increased resistance. Alternatively, the radiation itself may cure a cell by inducing the prophage and simultaneously inactivating some critical vegetative function, while the bacterial chromosome must emerge free of lethal damage.

Indirect induction. E. Borek and A. Ryan discovered in 1958 that unirradiated λ-lysogenic female (F^-) cells of *E. coli* can be induced by conjugation with UV-irradiated nonlysogenic F^+ cells. This effect, termed *indirect UV induction*, is not observed after treatment of the F^+ cell with other prophage-inducing agents such as X-rays or mitomycin C. Devoret and George (1967) showed that the polarity pattern for indirect induction resembles that for conjugational transfer: Lysogenic F^- cells are induced by interaction with UV-irradiated F^+ (or *colI*$^+$) cells, but lysogenic F^+ cells conjugating with UV-irradiated F^- cells are not induced. Thus, indirect induction apparently requires the transfer of UV-irradiated *episomal* DNA. However, conjugational transfer of chromosomal DNA from UV-irradiated *Hfr* cells (whose F factor is integrated into the chromosome) does not induce lysogenic F^- cells.

The extent of indirect induction is reduced by photoreactivating treatment of the UV-irradiated cell. Similarly, indirect induction by UV-irradiated wild-type cells is considerably less extensive than by excision-repair deficent host cells. Both facts together make it most likely that the photoproducts on the irradiated episome responsible for indirect lysogenic induction are pyrimidine dimers. Repair of these photoproducts can explain the relatively rapid loss of the irradiated episome's property of promoting indirect induction, whose half-life is approximately 8 minutes at 37°C. The present state of knowledge does not permit us to say whether or not indirect and direct UV induction of lysogenic cells are based on the same mechanism. The common requirement is the presence of UV-irradiated DNA in some form in the lysogenic cell, even though it is not clear why, in the case of indirect induction, only episomal DNA is effective. Perhaps the possible tying-up of existing prophage repressor molecules by some feature of irradiated DNA is important, in addition to delays in producing new repressor.

10.5 UV effects on genetic recombination

Multiple infection of host cells with phages carrying different genetic markers may result in the formation of recombinant progeny particles. Their frequency is increased if the parental particles had been UV-irradiated prior to infection; this observation holds for single recombinants, as well as for double recombinants obtained in three-factor crosses. Typically, the increase is proportional to the UV fluence until it reaches a limiting value, as illustrated in

Figure 10.11. The effect is reduced by photoreactivating treatment of the infected cells, suggesting that pyrimidine dimers in the phage DNA cause the increased number of recombination events.

The observed frequency of recombinants among progeny (N_{rec}/N) can be expressed formally by

$$\frac{N_{rec}}{N} = \frac{C_{rec}}{C} \times \frac{B_{rec}}{B}$$

where C_{rec}/C is the fraction of recombinant-producing complexes among the total complexes, and B_{rec}/B is the average burst size of recombinants *within recombinant-producing complexes* relative to the average burst size of all phages in all complexes. Thus, any enhanced frequency of recombinants after UV irradiation could result from an increase in either C_{rec}/C, or B_{rec}/B, or both. Techniques permitting selection for certain types of recombinants (see Section 7.8) can be applied in order to distinguish between these possibilities.

The increased frequency of recombinants after UV irradiation of the parental phage can be explained by two alternative hypotheses: (1) The presence of UV photoproducts in DNA stimulates recombination events; (2) the presence of UV photoproducts in DNA requires that progeny-producing complexes (survivors) undergo repair that involves recombination. The two hypotheses do not necessarily exclude each other, as we will see, but particularly strong arguments can be made in favor of the second hypothesis on purely logical grounds.

Genetic recombination requires infection of a host cell by two or more differently marked parental phages; therefore, if these particles are UV-irradiated, almost the entire progeny owe their existence to the occurrence of multiplicity reactivation (Section 7.7). This leads necessarily to a *selection in favor of recombinants* because the mechanism underlying multiplicity reactivation involves DNA recombination. Thus, we may conclude that the requirement for survival can only be satisfied by an increased number of recombination events, so that the increased frequency of any particular type of recombinant is an inevitable corollary.

As mentioned in Section 7.7, Epstein (1958) found that within *individual* bursts of multiplicity-reactivated complexes the fraction of recombinants (B_{rec}/B) was much higher than in complexes involving unirradiated parental phages. The predominance of one particular recombinant type of the two possible ones suggests that suitable recombination events are required *prior* to replication in order to provide for viability. Whether or not in addition the ratio C_{rec}/C is increased can only be decided in crosses between closely linked markers because otherwise the ratio is close to 1.0 even for unirradiated phages.

The question remains of whether in addition to hypothesis (2) there may be support for hypothesis (1). An experimental decision would require exclusion

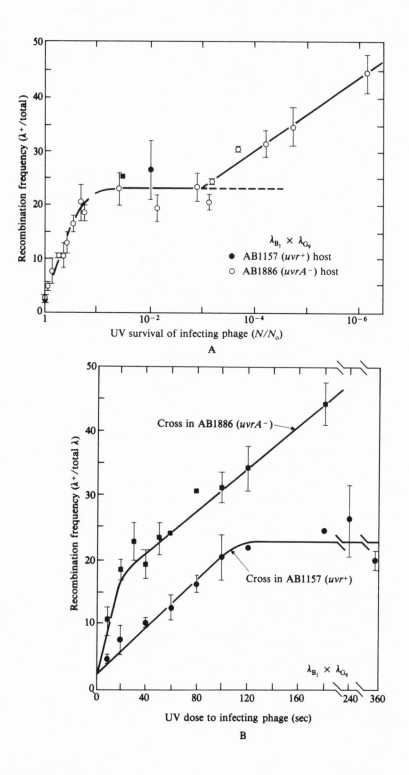

of multiplicity reactivation (and other recombination-related repair processes) without excluding at the same time genetic recombination itself. This is hardly feasible except for very small phages, like S13, containing only 6000 nucleotides in the form of single-stranded DNA. They show no measurable multiplicity reactivation, and their recombination frequencies are of the order of 10^{-3} to 10^{-4}. The fact that their low recombination frequency is considerably increased after UV irradiation suggests that photoproducts can enhance the intrinsic recombination tendency of DNA, at least in single-stranded DNA phages. Whether or not this may contribute in cases of larger phages with substantial multiplicity reactivation must remain an open question.

R. M. Baker and R. H. Haynes compared recombination frequencies in UV-irradiated phage lambda infecting either excision-repair proficient or deficient host strains. As panel A of Figure 10.11 indicates, recombination frequencies are identical in the two hosts if they are plotted as a function of survival of singly infecting phage. In contrast, if plotted as a function of the absolute UV fluence (panel B), recombination frequencies in the repair-deficient host cells are considerably higher than in repair-proficient host cells. This indicates that photoproducts responsible for enhanced recombination are as excision-repairable as lethal photoproducts are, which is not surprising if they are pyrimidine dimers. Moreover, the results show that excision repair itself does not enhance the likelihood for recombinant formation. To the contrary, those photoproducts remaining after excision repair or photoenzymatic repair cause the increased recombinant frequency because they must be eliminated by multiplicity reactivation or another recombination-related mechanism in order to produce viable progeny.

The results obtained with phages make it likely that increased recombination frequencies in such cellular organisms as bacteria, yeast, and other fungi are also a consequence of recombination-related repair processes. Genetic recombination is a dispensable luxury for unirradiated DNA, but becomes indispensable in some repair mechanisms guaranteeing survival. The inevitable selection in favor of recombinants is probably the basis for what we consider a UV-induced increase of recombination frequency.

In bacteria, recombinant formation is enhanced when UV-irradiated *Hfr* cells[1] of *E. coli* K12 cells conjugate with F$^-$ cells. Recombinants selected for relatively distant markers show an increased number of recombinants for unselected markers located between them. In other words: By recombination

Figure 10.11. Recombinant formation between UV-irradiated lambda phages in genetic crosses carried out in either excision-repair proficient (AB1157 *uvr*$^+$) or excision-repair deficient (AB1886 *uvrA*$^-$) host cells. *Panel A* shows that, for any given survival level, the recombinant frequencies obtained in the two types of host cells are identical. However, at any given absolute UV fluence, recombinant frequencies are considerably higher in phages infecting repair-deficient host cells (*Panel B*). (From R. M. Baker and R. H. Haynes, *Molec. Gen. Genetics 100*, 166, 1967.)

with the recipient chromosome, linked alleles on the transferred irradiated *Hfr* chromosome are separated more often than the same markers on an *unirradiated Hfr* chromosome. The same is true when unirradiated *Hfr* cells transfer chromosomal material into F⁻ cells, and the zygotes are subsequently irradiated. The enhancement of genetic recombination is much smaller, though still significant, if only recipient cells were irradiated prior to conjugation. In any case, exposure of the irradiated cells to photoreactivating light prior to conjugation greatly reduces this effect, indicating that pyrimidine dimers in DNA are the cause of increased recombinant formation.

In crosses between F⁺ males and F⁻ recipients the frequency of recombinants is likewise increased. In this case it probably represents enhanced integration of the F plasmid into the chromosome, a process requiring incubation of the irradiated F⁺ cells in broth for about one hour before mating. A similar incubation in minimal medium (supplemented to support growth of the auxotrophic cells) is ineffective. Many of these experiments were carried out at a time when little was known about repair processes, which makes their interpretation in terms of present knowledge speculative. It is reasonable to suggest that elimination of lethal photoproducts by recombination repair of necessity increases the recombinant frequency among the survivors. In addition, irradiated DNA in the course of such repair could temporarily assume a state in which its tendency to recombine is increased.

10.6 Inactivation of the cells' capacity to propagate phages

UV irradiation of bacterial cells destroys their ability to serve as hosts for *un*irradiated phages, a phenomenon called loss of *capacity*. The capacity of cells is usually much more UV-resistant than their colony-forming ability; but, within a given strain, the loss of capacity as a function of UV fluence varies widely for different types of phage. Not surprisingly, the capacity is more UV-sensitive the more a phage depends on the cell's metabolic functions. It is most sensitive for very small phages like ΦX 174, which contain only a few genes, and most resistant for such self-sufficient phages as the T-even group. Lambda, T1, T3, and others are intermediate in this respect.

Evidently loss of capacity can be the result of UV damage to certain cell components or inactivation of specific gene functions required for phage multiplication. The UV-damaged bacterial genome may also tie up, for its own repair, certain cellular enzymes that are needed for the normal propagation of undamaged phage. The latter would explain why the cells' capacity for phage lambda and others is more UV-resistant in excision-deficient (*uvrA*) strains, and much more UV-sensitive in *polA* strains than in wildtype *E. coli*. Presumably the lack of an early repair step (as in *uvrA*) precludes the involvement of other repair enzymes, whereas in the absence of DNA polymerase I excision repair continues in a different way that involves enzymes important for phage propagation. Accordingly, one finds that the capacity loss in *polA* cells is photoreversible, and that it is greatly diminished in the presence of excision-repair inhibitors.

Back in 1952, Dulbecco and Weigle discovered that, in strong contrast to far-UV radiation, near-UV and adjacent visible light destroys the capacity of bacterial cells more rapidly than it inactivates their colony-forming ability. The molecular events causing this rapid capacity loss are almost certainly different from those at far-UV wavelengths. An action spectrum might give indications as to whether or not there is a relationship to the photochemical processes underlying growth delay (Section 10.3.4).

11 Biological effects of solar UV radiation

11.1 Sunlight as a natural UV source

Sunlight is the basis for all life on earth. Its beneficial effects are well known and widely acknowledged, but little attention has so far been paid to the *adverse* effects of solar radiation. Apparently, it is hard to conceive that a ubiquitous, natural environmental factor, which has been in existence for more than a billion years of organismic evolution, could at the same time be harmful to life. Sunburn of the human skin and the inactivation of microbial cells by solar radiation are well-investigated, familiar phenomena, and the causal relationship between sunlight and skin cancer, suspected for more than half a century, has now been established beyond doubt. However, full recognition of the seriousness of potentially destructive effects of solar radiation comes only slowly with experimental evidence that the *observed* effects are only the tip of the iceberg. The vast majority of the *potential* damage is normally eliminated by cellular repair.

In our natural environment, solar UV is the only radiation with substantially harmful consequences on biological systems: the effects of ionizing radiations, either from the sky or from radioactive decay in the earth's crust, are comparatively small. Therefore, biological data obtained with sunlight are of vital interest for a quantitative assessment of its potential and actual damage. Studies with shorter UV wavelengths in the laboratory have proved most fruitful for understanding the sunlight effects because the major DNA photoproducts and their removal by repair processes are similar; nevertheless, such model studies cannot forever substitute for the real situation.

Electromagnetic radiation emitted by the sun ranges from radio waves to ultraviolet light and soft X-rays. Soft X-rays and vacuum UV are completely absorbed by air, and most of the far-UV radiation is absorbed by the stratospheric ozone; as a consequence, the shortest wavelengths reaching the earth's surface are slightly below 300 nm. The narrow overlap with the tail end of measurable nucleic acid absorption, extending to approximately 320 nm, accounts for the great importance of this spectral region with respect to adverse biological effects of sunlight. The incidence of solar energy to the earth's surface, within this critical region, depends strongly on such factors as the zenith angle of the sun, the thickness of the ozone layer,[1] cloud cover, altitude above sea level, local air pollution, and so on.

Variations in these parameters cause large fluctuations of the fluence rates

Table 11.1. *Relative effectiveness of photons in the 290–320 nm spectral region for altering DNA*

Wavelength (nm)	Relative effectiveness on DNA
320	0.03
315	0.1
310	0.6
305	2.6
300	15
295	60
290	160
For comparison:	
260	1000

Source: Data are taken from *Halocarbons: Environmental Effects of Chlorofluoromethane Release.* (A report of the Committee on Impacts of Stratospheric Change.) National Academy of Sciences, Washington, D.C., 1976.

at biologically relevant wavelengths, which are only a rather small fraction of the total solar emission. Physical measurements are not easy; to be biologically meaningful, they must be made for rather narrow wavelength bands below 320 nm because the relative biological effectiveness increases dramatically from 320 nm to 295 nm, as indicated in Table 11.1. Thus, the sun is a rather inconvenient UV source for experimental work, and one can hardly expect the results to match the reproducibility of laboratory experiments employing a technical UV source.

Lack of constancy of parameters in experimental work with sunlight makes a standardized measurement of the *biological efficiency* of a solar radiation in any experiment desirable. A UV-sensitive bacterial system, suitable for this purpose, is discussed in Section 11.2. Where only *relative* effects are of interest, they can be determined by comparison of one biological system with another in a given experiment under exactly identical conditions, whatever they are. Surprisingly good reproducibility is often obtained under stable weather conditions if a certain type of experiment is repeated several times within a few days at the same hour.

The high effectiveness of wavelengths around 300 nm requires that experimental samples in a petri dish or watchglass be either openly exposed to sunlight or covered with material transparent for this spectral region. Quartz plates are appropriate covers, whereas many of the commercial glasses are opaque or only partly transparent. For comparative purposes, filter glasses or filter liquids cutting off all or part of the most effective wavelengths can

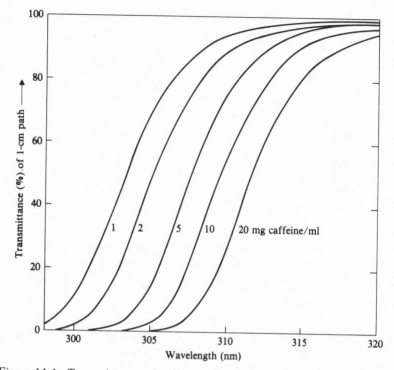

Figure 11.1. Transmittance of caffeine solutions. These solutions can be used as liquid filters displaying a sharp cutoff in the wavelength range 300–315 nm.

be employed. Aqueous caffeine solutions at various concentrations are suitable filter liquids, as their transmittances change in the critical region fairly sharply from complete opaqueness to almost full transparency (Figure 11.1). Such solutions can be conveniently used by filling them into a space between two parallel quartz plates, held at 1 cm distance by a frame of Plexiglas.

It is worth mentioning that sunlight also contains an abundance of wavelengths promoting photoenzymatic repair. Thus, photoreactivable biological systems exposed to sunlight display inactivation (or other adverse effects) and photorepair concurrently.

11.2 Solar UV effects on microbes, viruses, and DNA

Bacteria. The power of sunlight to inactivate bacteria was discovered a century ago by Downes and Blunt (1877). But only the past two or three decades of photobiological research have revealed the molecular processes underlying these germicidal effects and their modification by repair.

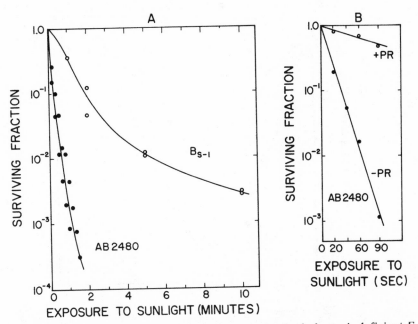

Figure 11.2. *Panel A*: Inactivation of the completely dark-repair deficient *E. coli* strain AB 2480 and of another sensitive strain, B_{s-1}, as a result of exposure to sunlight. *Panel B*: Survival of sunlight-exposed AB2480 cells that were either kept dark (lower curve) or exposed to photoreactivating, white fluorescent light (upper curve) before plating. (From W. Harm, *Radiation Res. 40*, 63, 1969.)

The potential extent of damage is best demonstrated by exposure of completely dark-repair deficient *E. coli* cells to solar radiation. The result in Figure 11.2A, obtained with a $uvrA^-$ $recA^-$ strain (AB 2480), shows that the number of viable cells can decrease by two to three orders of magnitude within just 1 minute. Cooling of the cell suspensions with ice water during the short exposure precludes most of the photoreactivation effect, but post-treatment solely with photoreactivating light demonstrates that about 90 percent of the damage is photorepairable, as illustrated in Figure 11.2B.

Such a biological system can be utilized as a biological dosimeter to monitor, at any time, the relative strength of germicidal solar radiation. The high sensitivity of cells is compatible with rather short exposure and permits an assessment of germicidal effects even when the sun is hidden behind a thick cloud cover. Because these cells reveal virtually *all* of the primary lethal damage, they are more suitable for consistent measurements than cells repairing a large fraction of the lesions. An example of dosimetric measurements on a sunny day in the fall at half-hour intervals is presented in Figure

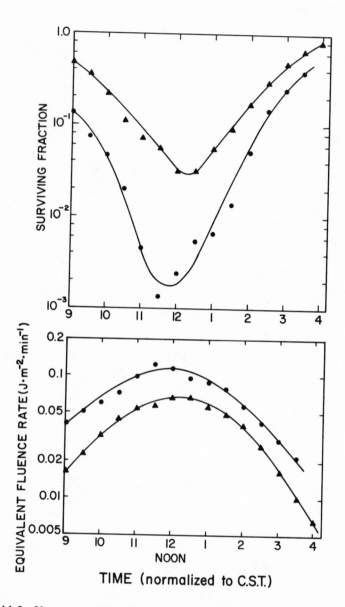

Figure 11.3. *Upper panel*: Survival of AB 2480 cells after 1-min sunlight exposure at different daytimes in late October (•) and early December (▲). *Lower panel*: Fluence rates, expressed as the equivalent fluence rates (in $J \cdot m^{-2} \cdot min^{-1}$) of 254-nm radiation, calculated from the values shown in the upper panel. (From W. Harm, *Radiation Res. 40*, 63, 1969.)

11.3, where the survival resulting from 1-minute exposures is plotted versus the daytime. The *germicidal effectiveness* is expressed by the fluence rate of 254-nm radiation that would be required for an identical effect. We notice that in this example the fluence-rate equivalents recorded are maximally $0.1 \text{ J} \cdot \text{m}^{-2}$ per minute, but still higher values can be obtained on other days.

The high sensitivity of strain AB 2480 emphasizes the importance of repair processes for the survival of cells exposed to sunlight. Survival curves of less repair-deficient strains or wildtype show decreased sensitivities, but in comparison with survival curves obtained at 254 nm, they reveal too high a sensitivity to sunlight. As an example: For inactivation to the 10^{-3} survival level, a nonphotoreactivable derivative of *E. coli* B/r requires a 600 times greater fluence of 254-nm radiation, but only a 30-fold longer exposure to sunlight, than AB 2480.

Thus solar UV damage is evidently less effectively dark-repaired than 254-nm UV damage. This suggests that sunlight produces, besides pyrimidine dimers, poorly repairable or nonrepairable lethal photoproducts, which are not or are rarely formed by 254-nm radiation. By filtering out the shorter wavelength region of the solar spectrum, one can show that such damage is mainly formed by wavelengths >360 nm. If dark- and photorepair processes combined would eliminate in *E. coli* wildtype cells virtually all of the sunlight-induced pyrimidine dimers, the relevance of such nondimer photoproducts for lethality would be obvious. The situation may indeed resemble that in *Micrococcus radiodurans* after 254-nm irradiation, where pyrimidine dimers are so extensively dark-repaired that the action spectrum indicates protein absorption as the cause of eventual lethality.

Free DNA and viruses. As one would expect, pyrimidine dimers are also the predominant, potentially lethal photoproducts after sunlight exposure of such noncellular biological systems as bacterial transforming DNA or phages. Transforming DNA of *Haemophilus influenzae* is readily inactivated by sunlight, as seen in Figure 11.4. This system was first used by C. S. Rupert in 1960 to demonstrate the high effectiveness of photoenzymatic repair in abolishing solar UV damage. If photoreactivating enzyme is added to a sunlight-inactivated DNA solution and the exposure continued, the transforming activity rapidly increases at a rate depending on the enzyme concentration, until the repair rate of lesions becomes slower than their formation rate (Figure 11.4A and B). On the other hand, in the presence of photoreactivating enzyme from the beginning of the exposure, the inactivation of transforming DNA proceeds at a reduced rate, representing the difference between the formation and repair of lesions (Figure 11.4C). As mentioned previously for sunlight-exposed cells, solar radiation damage to transforming DNA is also less effectively dark-repaired than 254-nm damage.

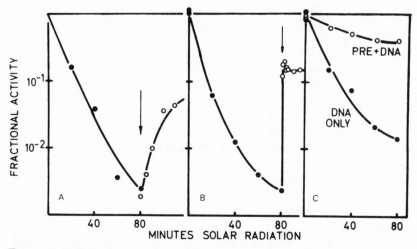

Figure 11.4. Solar inactivation of *Haemophilus* transforming DNA, and its repair in sunlight by yeast photoreactivating enzyme. Photoreactivating enzyme (PRE) was added after 80 min of sunlight exposure (arrow) either at low concentration (*panel A*) or at high concentration (*panel B*), or was present throughout the period of sunlight exposure (*panel C*). (From C. S. Rupert and W. Harm, in: *Advances in Radiation Biology*, L. G. Augenstein, R. Mason, and M. R. Zelle, eds., Vol. 2, Academic Press, New York, 1966, pp. 1–81.)

This is evident if one compares the sensitivities of transforming DNA to both types of radiation with those of completely repair-deficient *E. coli* cells.

The relative sensitivities of phage T4 wildtype and the mutants T4x^-, T4v^-, and T4v^-x^- are similar (though not identical) for solar UV and for 254-nm radiation (Figure 11.5). This indicates that the apparently dimer-specific v-gene repair system (see Section 7.5), and the x,y-gene repair system as well, eliminate solar UV damage at least as effectively as they remove lesions inflicted by shorter wavelengths. Furthermore, in Figure 11.6, the survival curves of sunlight-exposed phage T1, infecting either excision-repair proficient or deficient host cells, display differences similar to those after 254-nm irradiation of the phage.

Both transforming DNA and phages are nonmetabolizing systems of considerable stability and would thus appear suited for dosimetric use. Unfortunately, however, the small DNA target makes them fairly insensitive so that dosimetric determinations would require extended periods of exposure.

11.3 The importance of repair under natural conditions

We have seen that solar ultraviolet – the only radiation in our natural environment whose adverse effects on biological matter are substantial – forms

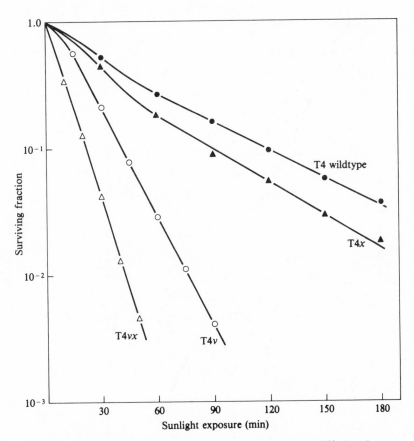

Figure 11.5. Inactivation by sunlight of T4 phage stocks differing in two phage-specific UV repair systems (see Sections 7.5 and 7.10). (From W. Harm, unpublished data.)

potentially lethal photoproducts in DNA at a considerable rate. One can infer, from the extraordinary sensitivity of completely dark-repair deficient cells, that microbial life in the open air would not be possible without a drastic reduction in the amount of lethal damage by repair processes. Consequently, we can assume that the short wavelength component of sunlight acts as a continual selective force for maintaining UV-related repair systems in many organisms.

The endangerment of life by solar radiation must have been even far more serious earlier in the earth's history, when the much lower oxygen content of the atmosphere provided little ozone, which now acts as a shield against the far-UV radiation. For the then prevailing protobiontic or unicellular forms of life that existed in water (which is essentially transparent for far UV),

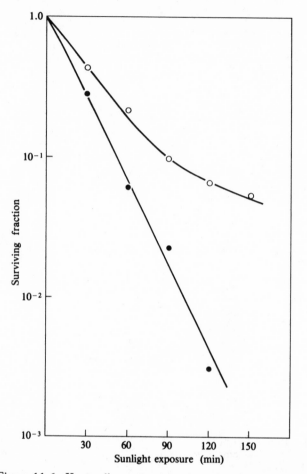

Figure 11.6. Host-cell reactivation of sunlight-inactivated phage T1. Samples of a sunlight-exposure phage suspension were plated either with an *E. coli* wildtype strain (upper curve) or with a strain defective in excision repair (lower curve). (From W. Harm, unpublished data.)

development of UV-repair mechanisms must have been a major step toward their expansion and further evolution. It is tempting to speculate that photo-enzymatic repair was perhaps the prototype of repair, not only because of its UV-specificity and utilization of solar energy itself, but also because its mechanism is so simple compared with the highly sophisticated, multi-enzymatic dark-repair systems.

Even now, with only a small fraction of destructive solar radiation reaching the earth, extensive repair of *some* kind is certainly a prerequisite for the life of most organisms. Although quite a number of organisms lack photoenzymatic repair, available data give no indications of the absence of either excision repair or postreplication repair in any well-investigated case. Because these repair systems also eliminate other kinds of damage besides UV lesions, their greater potential, both quantitatively and qualitatively, has perhaps made photorepair obsolete under certain circumstances. However, it is difficult to accept the occasional conjecture that photoreactivating enzyme must have some unknown function other than UV repair for the reason that *E. coli* cells living in the intestinal tract of mammals are never exposed to solar radiation. As a matter of fact, newborn mammals are free of colibacteria, and those by which they are subsequently infected may have been extensively exposed to sunlight outside the body.

Loss of a repair function due to mutation or temporary inhibition can have severe consequences for the individual. An example is the human hereditary disease xeroderma pigmentosum, which manifests itself as an abnormal sensitivity of the skin to sunlight, resulting in multiple skin malignancies with mostly fatal consequences. It was recently discovered that cells of xeroderma patients are unable to excise pyrimidine dimers, and subsequent studies have indicated the importance of repair systems for some other diseases also. Because their characteristics are closely associated with the problem of UV carcinogenesis, they are discussed in greater detail in Chapter 12.

11.4 Effects of sunlight on the human skin

Interaction of sunlight with the human body can have both beneficial and harmful consequences. Sunlight generally supports the body in combating infectious diseases, and it provides for the conversion of provitamin D (ergosterol) into vitamins D_2 and D_3. Insufficient sunlight during a child's development used to be the cause of rickets, a now rare disease thanks to healthier living conditions and/or addition of D vitamins to milk. These unquestionably desired effects of sunlight contrast with various adverse reactions of the skin, particularly after *excessive* exposure to wavelengths at the short end of the solar spectrum. Presumably owing to the widespread cultural addiction to suntan, some of the serious consequences of overexposure to sunlight are now found at increasing frequencies. They have long been the concern of dermatologists, but are neglected by many other people.

The most common adverse effect of solar UV on the human skin is the *erythema* (sunburn), a reddening of the skin, resulting from dilation of the blood capillaries, with subsequent blistering and peeling. In severe cases the symptoms resemble those of burns, and the outcome can be fatal. Wavelengths below 320 nm are responsible for the erythematous reaction, the shortest ones being the most effective. For this reason, inexperienced persons often suffer from severe sunburn at high altitudes, or in geographic areas with high insolation and dry, clean air. The effects are intensified by reflection from snow, water, or light-colored sand. Protective action of ointments against sunburn is usually achieved by their strong absorption below 320 nm; transparency for longer wavelengths still permits the cosmetically desired tanning of the skin.

The primary photochemical processes underlying sunburn are not well understood; apparently they occur in the epidermis. Erythema caused by still shorter wavelengths (not present in sunlight), with an efficiency maximum around 250 nm, is presumably due to reactions in the dermis. A person's amenability to erythematous skin reactions depends largely upon the skin pigmentation; as a rule, lighter skin is more likely to suffer from solar radiation, but individual differences, even among people with similar skin pigmentation, are considerable.

Another acute reaction to short wavelengths from sunlight, and more so from technical UV sources (e.g., 254 nm), is *conjunctivitis*. It is an inflammation of the mucous membrane that extends from the inner surface of the eyelid to the forepart of the eyeball. This harmful effect can be easily avoided by wearing appropriate eyeglasses or goggles.

Long-term exposure to solar radiation, which is characteristic for outdoor workers and certain occupations (farming, fishing, etc.), produces, with progressing age, the symptoms of *connective elastosis* of the skin ("sailor's skin"). This is, in essence, a degeneration of the collagen and the elastic connective tissue of the dermis. Typically, those parts of the body are affected that are least protected against the sun, as hands and forearms.

Among the most serious consequences of solar irradiation are *neoplasms* of the skin. A causal relationship between sunlight exposure and the incidence of skin cancer has been suggested since the turn of this century, and evidence for it has accumulated ever since then (see Chapter 12 for a more detailed discussion). Sailors and farmers have a considerably higher incidence of skin cancer than office employees. A study performed in many areas of the United States clearly revealed a positive correlation between the development of skin cancer and the average amount of insolation. If the stratospheric content of ozone were substantially to diminish, the situation would undoubtedly worsen, threatening not only human health, but life on earth in general. Concern about this matter prompted recent U.S. government-sponsored studies to determine the impact of supersonic and other high-altitude air traffic, as well as of halogenated hydrocarbons (used as spray can propellants or refrigerants) on the ozone layer. The results are still controversial; in particular, the evaluation of measurements encounters the difficulty that even without human interference the stratospheric ozone content varies in a complex manner, as O_3 molecules are continuously formed and destroyed by cosmic processes.

12 UV carcinogenesis

12.1 General

Over the past several decades, evidence has accumulated indicating a strong correlation between the carcinogenicity and mutagenicity of many chemical and physical agents. As a consequence, inexpensive and fast mutagenicity tests with microorganisms are now widely used to obtain indications of the possible carcinogenicity of an agent. Because mutagenic alterations of necessity occur in DNA, one can expect, if the mentioned correlation holds, that alterations leading to the transformation of a normal cell into a cancer cell likewise occur in DNA.

Carcinogenicity of the UV component of sunlight has been suspected since the turn of the century, and experimental evidence for it has existed for about as long as the mutagenic effects of UV radiation have been known. More recent experimental work, particularly in conjunction with repair of carcinogenic lesions, has shown beyond doubt that the basic photoproducts occur in DNA, and are similar in nature to those causing UV-induced mutability and UV inactivation.

Considering that the causal relationship between UV radiation and carcinogenesis is fairly well established, the main thrust of research is now toward understanding the *mechanism* of UV carcinogenesis both at the molecular and at the cellular level, and toward assessment of the carcinogenic potential of sunlight upon the human skin. Studies of UV carcinogenesis may provide models for understanding the action mechanisms of other carcinogenic agents and their repair, much as the lethal and nonlethal effects of UV on microorganisms have been models for the investigation of effects of other physical and chemical agents, and much as the repair of UV photoproducts has helped our understanding of the repair and correction of other DNA alterations.

As far as UV carcinogenesis in the human skin is concerned, most of our knowledge is based on epidemiological studies rather than experimental work. The latter, if we disregarded the impropriety of experimental studies with human beings, would also be difficult because of the extensive periods (usually several years) required for tumor induction in men. Thus experimental evidence for UV carcinogenesis comes mainly from studies with either laboratory mammals or cell cultures of human or other mammalian origin. For studies on whole animals, rats and mice are frequently used because they develop tumors within short time periods. The availability of albino strains and hairless strains provides an additional advantage, as the UV radiation can reach and penetrate their skin more easily than the skin of wildtype animals. The use of mammalian cell cultures has particularly contributed toward an understanding of the molecular basis of carcinogenesis.

12.2 Evidence relating human skin cancer to UV exposure

Relevant facts relating sunlight exposure and human skin cancer are these: (1) Human races with light skin are much more susceptible to skin cancer than heavily pigmented races; (2) in light-skinned persons the frequency of skin cancer increases with the amount of sunlight typical of the geographic area in which they live; (3) persons spending much of their time outdoors have a higher incidence of skin cancer than people staying mostly indoors; (4) skin cancers develop predominantly on the sun-exposed parts of the body. Although these correlations do not constitute proof of a causal relationship between sunlight exposure and incidence of skin cancer, they are nevertheless strongly indicative in this regard. Because they are supported by experimental results with laboratory animals, we can consider them a reasonable basis for our current thinking.

At least three types of skin cancer result from sunlight exposure: *basal cell epitheliomas, squamous cell carcinomas*, and *melanomas* of several kinds. The first two types are the most common, but fortunately their tendency to metastasize is small. In contrast, the rapidly growing melanomas metastasize easily and cause a high mortality rate. In the United States, the average annual incidence of malignant melanoma in recent years was 4.1 to 4.4 per 100,000 among white persons, whereas it was only 0.6 to 0.7 per 100,000 among black people. Apparently, the dark pigmentation of the skin provides substantial protection.

Figure 12.1 shows the correlation of the mortality rate due to skin melanoma in the United States with the geographical latitude, which suggests solar UV radiation as the cause. This view is supported by fact that the most sun-exposed parts of the body, in particular the nose, lips, forehead, and back of the neck, are those most likely to develop skin cancers. This was long ago established for carcinomas and epitheliomas, but was not apparent for melanomas. However, extensive recent studies of a large number of melanoma cases showed that they are conspicuously rare or absent in those areas of the body usually covered by bathing suits. Figure 12.2 shows a composite picture of these cases, suggesting the potential formation of melanoma as a consequence of *overexposure* to sunlight of usually lightly pigmented parts of the body, as typically occurs when sunbathing on the beach, around the swimming pool, and so forth.

Superimposed upon the general correlations found between insolation and skin cancer development are certainly large variations regarding cancer

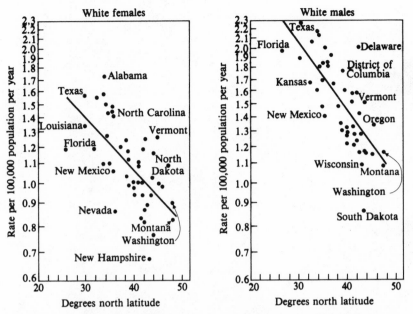

Figure 12.1. Annual mortality by skin melanoma in the United States as a function of geographical latitude for white females (*left panel*) and white males (*right panel*). (From *Measurements of Ultraviolet Radiation in the United States and Comparisons with Skin Cancer Data*; prepared by Scotto, Fears and Gori. Publ. by U.S. Dept. of Health, Education and Welfare, Public Health Service, National Institutes of Health, Nov. 1975.)

susceptibility between individuals of the same racial origin or external appearance. As a rule, persons who easily suffer sunburn, that is, those with light eyes, blond hair, and fair skin, are most likely to develop skin cancer upon extensive exposure to sunlight. Characteristically, a population that originated in Scotland (where people with reddish hair and very light skin pigmentation are rather common) and now lives in the tropical, northernmost part of Australia has become known to be extremely susceptible to skin neoplasms.

12.3 Evidence for pyrimidine dimers in DNA as a cause of neoplastic transformation

12.3.1 Range of wavelengths effective in carcinogenesis

UV wavelengths for which carcinogenic action is clearly established range from about 230 to 320 nm. Because we are dealing here with entire biological tissue, rather than with individually suspended cells, it is hardly possible to

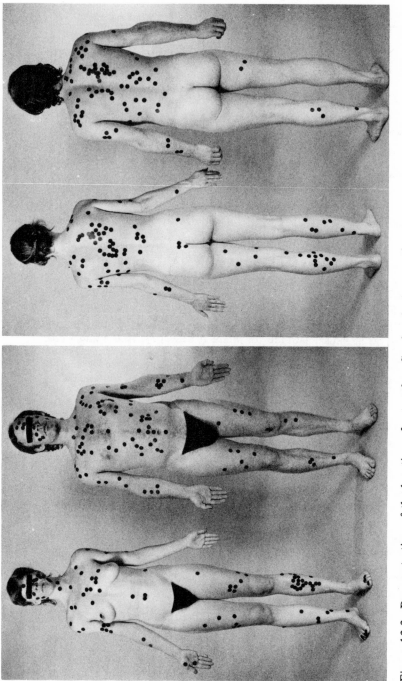

Figure 12.2. Demonstration of the location of more than five hundred cases of malignant skin melanomas on the female and male bodies. (Data from Drs. T. B. Fitzpatrick, A. J. Sober, B. J. Pearson, and R. Lew, compiled from the Melanoma Clinical Cooperative Group, comprising Harvard University, Temple University, New York University, and University of California at San Francisco. Courtesy of Dr. Fitzpatrick.)

determine the target for UV carcinogenesis by means of action spectra, as described in previous chapters for mutagenesis and inactivation of microbial cells or viruses. In those cases, the likelihood for photon absorption by an individual is small enough that the wavelength dependence of the effect matches the absorption by the target molecules. In contrast, a tissue of >1 mm thickness absorbs all ultraviolet radiation completely; only the average depth of penetration differs for different wavelengths.

Evidence to be presented in Section 12.3.2 strongly indicates that DNA is the primary target for carcinogenic action of ultraviolet radiation. The limitation of effectiveness to the 230-320 nm spectral region comes as no surprise: Greater wavelengths are not measurably absorbed by DNA, whereas wavelengths below 230 nm cannot reach the DNA target because of strong shielding by the cell proteins. On the other hand, the general experience that 280-300 nm radiation is more carcinogenic than 254-nm radiation is hardly an argument against DNA being the target; most likely, the tissue absorbing 254 nm more readily than 280-300 nm prevents the 254-nm radiation from reaching the most responsive cells.

12.3.2 Photoreversal of carcinogenic effects

Photoenzymatic repair is the only general repair mechanism so far known to abolish specifically pyrimidine dimers in DNA. Therefore, if photoreactivating treatment substantially reduces the incidence of neoplasms after UV irradiation, one can reasonably conclude that neoplastic transformation of cells can result from the presence of UV-induced pyrimidine dimers in their DNA. This thought was the basis for experiments by R. W. Hart and R. B. Setlow on the UV induction of granulomas and thyroid carcinomas in fish. The species *Poecilia formosa* is particularly suited to this kind of research, as it propagates gynogenetically (i.e., the offspring arise from nonfertilized eggs) so that a clone of genetically identical individuals can be obtained from a single female. Like other fish, *P. formosa* contains considerable amounts of photoreactivating enzyme.

Figure 12.3 schematically outlines the experimental procedure. Cell suspensions of tissue homogenate from one individual of the clone are treated in various ways, approximately $3\text{-}5 \times 10^5$ cells are injected into the back muscle or the abdominal cavity of other members of the clone. Exposure of the cells to 10-20 $J \cdot m^{-2}$ of 254-nm radiation results in development of granulomas in approximately 85 percent of the injected individuals, and of thyroid carcinomas in virtually all of them. Quite similar results are obtained with cells exposed to photoreactivating light (320-420 nm) *prior* to UV irradiation, whereas the photoreactivating light alone does not induce tumors. However, exposure to photoreactivating light *after* far-UV irradiation strongly reduces

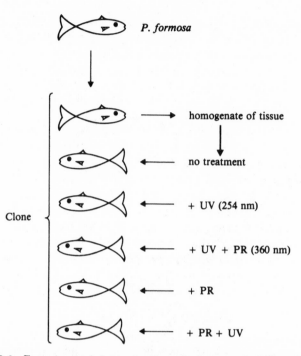

Figure 12.3. Experimental determination of photorepairability of neoplastic transformation by 254-nm UV radiation in *Poecilia formosa* (see text). (From R. W. Hart and R. B. Setlow, in: *Molecular Mechanisms for Repair of DNA*, Part B, P. C. Hanawalt and R. B. Setlow, eds., Plenum Publ. Co., New York, 1975, pp. 719-24.)

the fraction of individuals with either type of tumor, as well as the average number of tumors per fish.

The calculated photorepairable sectors in these experiments range from 0.6 to >0.9, suggesting that the tumors result overwhelmingly from the formation of pyrimidine dimers in DNA. Furthermore, it is estimated that a fluence of $5 \ J \cdot m^{-2}$, producing in each cell about 10^5 pyrimidine dimers, leads to an average of one transformed cell among 10^5 thyroid cells. Consequently, the probability for a dimer to cause neoplastic transformation must be of the order of 10^{-10}.

Skin neoplasms and UV-induced necroses in the ear of albino mice were observed after short-wavelength UV-irradiation by various authors more than 20 years ago. They found that subsequent application of photoreactivating light significantly reduced the frequency of these effects. Such results were regarded with caution, as there was no evidence for photoreactivating enzyme in cells of placental mammals (in contrast to marsupial mammals, which contain considerable amounts of this enzyme). However, the recent discovery of photoreactivating enzyme in cells of higher mammals and humans by B. Sutherland (1974) is consistent with these earlier observations, suggesting, like the fish experiments, that UV-induced skin malignancies are largely caused by pyrimidine dimers.

12.4 Dark repair of carcinogenic UV lesions

Another example of UV-induced malignancies of the skin, caused primarily by pyrimidine dimers in DNA, is the human hereditary disease xeroderma pigmentosum (see Sections 7.3 and 11.3). Although a rare disease because of its autosomal-recessive inheritance pattern, it was known even in the last century. Its characteristic is an unusual sensitivity of the individual to sunlight, resulting in pigmentation changes, hyperkeratosis, and development of multiple skin cancers. Recent genetic experiments, involving the cell fusion technique and complementation tests, have shown that mutations in at least five different genes (complementation groups) can cause the clinical phenotype of classic xeroderma. Genetically unrelated to xeroderma is the *De Sanctis-Cacchione syndrome*, a clinical form combining the xeroderma skin characteristics with neurological complications, growth retardation, and mental deficiencies.

The essential discovery establishing the molecular basis of xeroderma pigmentosum was made by Cleaver in 1968, who showed that cells from patients with this disease are defective in excision repair of UV-induced pyrimidine dimers. Additional studies indicated that the degree of repair deficiency varies for different complementation groups. In a few cases of the disease, usually called XP variant, pyrimidine dimers are excised normally, and postreplication repair is instead defective. Differences between various xeroderma complementation groups regarding their degree of excision-repair deficiency are also reflected by similar differences in the extent of host-cell reactivation of UV-irradiated DNA viruses (herpes viruses, SV40, adenoviruses, and others), which was illustrated in Figure 7.9. The repair defect in classic xeroderma pigmentosum cases and the De Sanctis-Cacchione syndrome is a lack of endonucleolytic cleavage of the DNA strand adjacent to a pyrimidine dimer, analogous to the repair deficiency in *E. coli* mutants of the *uvrA* or *uvrB* type.

Another rare human disease, displaying defective excision repair, is *Fanconi's anemia*. Like xeroderma, its inheritance pattern is autosomal-recessive, and it is characterized by a high incidence of cancer, chromosomal aberrations, and aplastic anemia. Although the DNA of UV-irradiated fibroblast cells from Fanconi's anemia patients contains strand scissions, indicating proficiency for the first step in excision repair, the scissions persist when normal cells have completed their repair. This suggests that in the case of this disease excision repair either is very slow or is defective in one of the later steps, for example, repair synthesis and/or rejoining.

There exist a number of other human diseases related to defective repair, although not necessarily repair after UV damage. Even though they are infrequent, they make us aware of the vital importance of repair processes as means of protection against cellular damage.

13 Sensitized UV effects on DNA

13.1 Substitution for thymine by 5-bromouracil

We have seen in previous chapters that most of the biological effects of UV radiation reflect the absorption properties of DNA. Though the ratio of (A + T)/(G + C) in double-stranded DNA from different organisms varies, their action spectra are not affected to a great extent. However, action spectra can change considerably if the absorption properties and the resulting photochemical effects are altered by substituting base analogues for natural DNA bases. The best-investigated example is 5'-bromouracil, which can be incorporated, by certain techniques, into DNA to replace virtually all the thymine.

The main differences in UV effects between bromouracil-substituted and unsubstituted DNA are these: (1) The absorption spectrum of substituted DNA is slightly shifted toward longer wavelengths. This shift hardly affects the absorbance in the peak region, but does so considerably at wavelengths above 300 nm, where the absorbance of substituted DNA is 10- to 100-fold higher than that of unsubstituted DNA. Therefore, in a mixture of substituted and unsubstituted DNA molecules (or of cells containing substituted and unsubstituted DNA) the inactivation by certain wavelengths (for example, the mercury line at 313 nm) is extremely selective. (2) The photoproducts in bromouracil-substituted DNA differ qualitatively and quantitatively from those in unsubstituted DNA. As a result, repair of lethal damage in substituted DNA is often much less effective than in natural DNA. Photoenzymatic repair, and v-gene repair of phage T4, both specific for pyrimidine dimers, are virtually absent. Most of the excision repair of bromouracil-substituted DNA is absent, as revealed by the low extent of host-cell reactivation in phage and the low survival of substituted bacterial cells themselves.

The major photoproduct in the substituted DNA is uracil, formed by debromination. Debromination affects the deoxyribose moiety of the phosphodiester chain, leading to single-strand breakage, perhaps upon formation of an alkali-labile bond. In hybrid DNA consisting of a substituted and an unsubstituted strand, most of the breakage is in the bromouracil-containing strand, but double-strand breakage can occur as a secondary effect. The fact that the number of double-strand breaks in such hybrid molecules increases linearly with the UV fluence suggests that they result from a single photon absorption.

DNA strand breakage is the major cause of lethality in bromouracil-substituted DNA. In the presence of *cysteamine*, which as a radical scavenger prevents hydrogen abstraction from the deoxyribose by the uracil radical, DNA strand breakage or alkali-labilization is much reduced. As a consequence, in substituted transforming DNA, or in phages containing substituted DNA, lethal UV lesions resulting from irradiation in the presence of cysteamine are repairable.

The selective sensitivity of bromouracil-substituted DNA to strand breakage by 310–320 nm radiation can be used as an experimental tool. For example, if the resynthesis step in excision repair is forced to utilize the substitute, subsequent exposure of the DNA to a sufficiently high fluence of 313-nm radiation permits determination of the number of single-strand breaks and thus the number of repaired regions. When this number is known, the rate of the occurrence of breaks per molecule with the UV fluence can be related to the average length of the resynthesized sequence. Detailed reviews of the photochemical effects in bromouracil-substituted DNA and their biological consequences are found in papers by Rahn and Patrick (1976) and Setlow and Setlow (1972).

Other base substitutions. 5-Iodouracil has similar effects but is used much less than 5-bromouracil. It can serve as a treatment of *Herpes* virus infections of the human cornea because the rapidly replicating virus DNA incorporates this analogue and is subsequently inactivated by light. In contrast, DNA whose thymine is substituted by 6-azathymine does not form UV photoproducts involving 6-azathymine. Apparently for this reason, cells with partly azathymine-substituted DNA show reduced UV sensitivity.

13.2 UV sensitization of DNA by exogenous molecules

13.2.1 Sensitized thymine dimerization

The singlet state of *acetophenone* has a lower energy level than the singlet states of the four DNA bases; therefore, absorption at wavelength >300 nm is appreciable. Excited acetophenone converts with high efficiency to its triplet state, whose energy level is higher than the triplet state of thymine, but not of any of the other bases. As shown by Lamola and Yamane in 1967, energy transfer from triplet acetophenone to thymine can occur by collision, resulting in thymine-thymine dimerization. Because virtually no other pyrimidine dimers are formed under these conditions, it was hoped that acetophenone sensitization would permit a distinction between the effects of thy◇thy dimers versus the other types of pyr◇pyr dimers in biological materials. But the occurrence of additional strand breaks and the oxygen dependence of the dimerization/strand breaks ratio has made the interpretation of biological data difficult.

Nevertheless, acetophenone sensitization at 313 nm has the advantage that it permits an unusually high percentage of thymine (up to 37 percent) to become dimerized. Because monomerization of dimers at this wavelength is negligible, the theoretical maximum of dimerization in *E. coli* DNA would be 47 percent of the thymine. The fact that this is achieved only by irradiation of *denatured* DNA in the presence of the sensitizer suggests that in native DNA certain constraints prevent complete photodimerization.

Other triplet sensitizers used experimentally are *acetone* and *benzophenone*. The triplet energy of acetone is sufficiently high to transfer it to either thymine or cytosine and thus cause formation of all three types of pyr ◇ pyr dimers usually observed in UV-irradiated DNA.

13.2.2 Psoralen

Psoralen belongs to the group of furocoumarins. Like some of its derivatives it is a photosensitizer to human and mammalian skin, which causes tanning and erythema by light of the 300–400 nm region. With respect to the inactivation of microbial cells, either psoralen or near UV alone is much less effective than the two together. A certain level of inactivation by 360-nm radiation requires only $\frac{1}{1000}$ of the fluence if psoralen is present. Unlike the effects of other photosensitizers, psoralen-mediated UV damage does not depend on oxygen and shows a *negative* temperature response.

Psoralen-sensitized photo-effects in vitro as well as in vivo involve reactions with DNA; a major photoproduct of biological importance is a crosslink between the two complementary strands of native DNA. Such crosslinks, which are stable to alkali or heat, occur as a result of single-photon absorption, as the exponential decrease with UV fluence of the fraction of noncrosslinked DNA molecules indicates. Psoralen-induced crosslinks resemble those induced by mitomycin C or bifunctional alkylating agents, all of which are efficiently eliminated by dark-repair systems removing UV-induced pyrimidine dimers. In completely dark-repairless ($uvrA^-recA^-$) cells of *E. coli* a single crosslink is sufficient for inactivation. The natural occurrence of psoralens may well explain the extensive repair of psoralen-sensitized photodamage.

Irradiation with near UV in the presence of 8-methoxy psoralen is mutagenic for *E. coli* cells. Like pyrimidine dimers, the premutagenic lesions are excision-repaired; consequently repair-deficient cells show a greater mutagenic response to the sensitized near-UV treatment than repair-proficient cells. exr^- cells are not mutated at all; apparently the eventual mutagenic alteration in DNA involves an error-prone step in postreplication repair that applies equally to psoralen crosslinks and to pyrimidine dimers.

13.2.3 Desensitization of DNA to UV radiation

Binding of acridine dyes by DNA strongly reduces the rate of pyrimidine dimerization by far-UV irradiation. This desensitizing effect is already observable when the ratio [bound dye molecules]/[nucleotides] is 10^{-2} or less; therefore, it is probably caused by energy transfer from DNA to the dye, rather than by a structural change in the DNA molecule, due to intercalation of the dye into the double-helix. Proflavine, acridine orange, methyl green, ethidium bromide, and chloroquine have all been found effective in protecting the DNA against photodimerization.

As one would expect, desensitization of DNA by acridines for the formation of photoproducts likewise reduces the biological effects of far-UV irradiation. Inactivation of *E. coli* cells pretreated with 5 μg/ml acriflavine for 30 minutes and irradiated in the presence of the dye requires a 10 times greater UV fluence than the same extent of inactivation in the absence of the dye.

Notes

Chapter 1

1. 1 nm = 1 nanometer = 10^{-9} meter (m) or 10^{-7} centimeter (cm). Sometimes wavelengths are expressed in Ångström units (1 Å = 10^{-1} nm).

Chapter 2

1. Previously this quantity was called UV dose or incident UV dose. The change was recommended by a study performed under the auspices of the U.S. National Committee for Photobiology, because the term dose is already preempted by its use in connection with ionizing radiations, where it means the *energy absorbed* (see Rupert, 1974). For this reason the term fluence will be used throughout the book.
2. Watts per square meter, since 1 J = 1 Wsec.
3. Even if ϵ is the same and the number of photons absorbed is the same as in dilute solution, the photochemical reactions could still differ in the particle from those that occur in solution, owing to the different surroundings of the excited molecule.

Chapter 3

1. The quantum yield is the probability with which an *absorbed* quantum causes a particular, photochemically or photobiologically defined effect.
2. When cells of certain bacterial species, in particular of the genera *Bacillus* and *Clostridium*, approach the end of their growth period, they form inside their cytoplasm a "spore," which is afterwards released. Bacterial spores have extremely low water content and their DNA and other components required for later "germination," i.e., conversion to a vegetative cell, is enclosed by an assembly of heavy membranes. Spores are very resistant to dryness and heat, and are also less sensitive to UV and ionizing radiation than vegetative cells.

Chapter 4

1. In terms of the classical target theory the one-hit dose-effect curve is the complementary function $N/N_0 = 1 - (S/S_0) = 1 - e^{-cF}$, with N being the number of inactivated particles, and N_0 the total number of particles. However, because the survival expresses directly the experimental result, almost all recent publications show survival, rather than inactivation dose-effect curves. Also, because inactivation is usually studied at levels where the great majority of the population is affected, it is easier to deal with the survival function equation (4.1a), rather than with the complementary function.

2. The Poisson formula describes the *random* distribution of events within units, when the average number of events per unit is not high. In general terms, the probability $p(n)$ for exactly n events occurring within one unit, if the average number of events per unit is a, is expressed by

$$p(n) = \frac{a^n \cdot e^{-a}}{n!}$$

Thus $p(n = 0) = e^{-a}$,

$p(n = 1) = ae^{-a}$,

$p(n = 2) = (a^2/2) e^{-a}$, etc.

3. This is sometimes called D_{37}, or $D_{0.37}$, for the dose (now fluence) that leaves e^{-1} survivors. Replacing the old term dose by fluence, we will now call it F_{37} or $F_{0.37}$.
4. Actually the AT/GC base ratio in T5 DNA is greater than in the others, and should result in a higher Φ. This would make the differences in intrinsic sensitivities between T5 and the other phages still larger.

Chapter 7

1. We derive from equation (7.5) that the fraction of multicomplexes equals $1 - (me^{-m}/1 - e^{-m})$; for $m \ll 1$ this expression approaches $0.5\ m$.
2. By the *single burst technique* phage-infected cells are randomly distributed into many separate tubes at an average number $\ll 1$ per tube. Consequently, after lysis most tubes contain no phages, whereas those with phages represent in most cases the progeny from a single, infected cell.

Chapter 10

1. *Hfr* stands for high frequency of recombination, in contrast to the low frequency of recombination typical of $F^+ \times F^-$ crosses. Only an *Hfr* cell can transfer chromosomal material into the recipient (F^-) cell. The low chromosomal recombination observed in $F^+ \times F^-$ crosses is a consequence of integration of the F episome into the bacterial chromosome, which provides it with the *Hfr* property.

Chapter 11

1. The ozone (O_3) content of the upper atmosphere for a given geographical area can be expressed by the thickness (of the order of a few millimeters) of a layer that the O_3 molecules would occupy if their pressure were 1 atmosphere.

References

Adler, H. I., W. D. Fisher, A. A. Hardigree, and G. E. Stapleton (1966). Repair of radiation induced changes to the cell division mechanism of *Escherichia coli. J. Bacteriol. 91*, 737–42.

Altenburg, E. (1930). The effect of ultraviolet radiation on mutation. *Anat. Rec. 47*, 383.

Ananthaswamy, H. N., and A. Eisenstark (1976). Near-UV-induced breaks in phage DNA: Sensitization by hydrogen peroxide (a trytophan photoproduct). *Photochem. Photobiol. 24*, 439–42.

Avery, O. T., C. M. McLoed, and M. McCarty (1944). Studies on the nature of the substance inducing transformation of pneumococcal types. Induction of transformation by a desoxyribonucleic acid fraction isolated from pneumococcus type III. *J. Exper. Med. 79*, 137–58.

Borek, E., and A. Ryan (1958). The transfer of irradiation-elicited induction in a lysogenic organism. *Proc. Natl. Acad. Sci. U.S. 44*, 374–7.

Boyce, R. P., and P. Howard-Flanders (1964). Release of ultraviolet light-induced thymine dimers from DNA in *E. coli* K12. *Proc. Natl. Acad. Sci. U.S. 51*, 293–300.

Clark, A. J., and A. D. Margulies (1965). Isolation and characterization of recombination deficient mutants of *E. coli* K12. *Proc. Natl. Acad. Sci. U.S. 53*, 451–9.

Cleaver, J. E. (1968). Defective repair replication of DNA in xeroderma pigmentosum. *Nature 218*, 652–6.

Demerec, M., and R. Latarjet (1946). Mutations in bacteria induced by radiations. *Cold Spring Harbor Symp. Quant. Biol. 11*, 38–49.

Devoret, R., and J. George (1967). Induction indirecte du prophage λ par le rayonnement ultraviolet. *Mutation Res. 4*, 713–34.

Doudney, C. O., and C. S. Young (1962). Ultraviolet light induced mutation and deoxyribonucleic acid replication in bacteria. *Genetics 47*, 1125–38.

Downes, A., and T. P. Blunt (1877). Researches on the effect of light upon bacteria and other organisms. *Proc. Royal Soc. London 26*, 488–500.

Drake, J. W. (1973). The genetic control of spontaneous and induced mutation rates in bacteriophage T4. *Genetics Suppl. 73*, 45–63.

Dulbecco, R. (1949). Reactivation of ultraviolet inactivated bacteriophage by visible light. *Nature 163*, 949–50.

Dulbecco R., and J. J. Weigle (1952). Inhibition of bacteriophage development in bacteria illuminated with visible light. *Experientia 8*, 386.

Elkind, M. M., and G. F. Whitmore (1967). *The Radiobiology of Cultured Mammalian Cells*. Gordon & Breach, New York.

Ellis, E. L., and M. Delbrück (1939). The growth of bacteriophage. *J. Gen. Physiol. 22*, 365–84.

Epstein, R. H. (1958). A study of multiplicity-reactivation in the bacteriophage T4. I. Genetic and functional analysis of T4D-K12(λ) complexes. *Virology 6*, 382–404.

Errera, M. (1952). Etude photochemique de l'acide désoxyribonucléique. I. Mesures énergétiques. *Biochim. Biophys. Acta 8*, 30–7.

Friedberg, E. C. (1975). Dark repair in bacteriophage systems: Overview. In: *Molecular Mechanisms for Repair of DNA* (P. C. Hanawalt and R. B. Setlow, eds.), Part A. Plenum Press, New York, pp. 125–33.

Garen, A., and N. D. Zinder (1955). Radiological evidence for partial genetic homology between bacteriophage and host bacteria. *Virology 1*, 347–76.

Gates, F. L. (1928). On nuclear derivatives and the lethal action of ultraviolet light. *Science 68*, 479–80.

Harm, W. (1958). Zur Deutung der unterschiedlichen UV-Empfindlichkeit der Phagen T2 und T4. *Naturwissenschaften 45*, 391.

Harm, W. (1964). On the control of UV sensitivity of phage T4 by the gene *x*. *Mutation Res. 1*, 344–54.

Harm, W., H. Harm, and C. S. Rupert (1971). The study of photoenzymatic repair of UV lesions in DNA by flash photolysis. In: *Photophysiology* (A. C. Giese, ed.), Vol. 6. Academic Press, New York, pp. 279–324.

Haynes, R. H. (1966). The interpretation of microbial inactivation and recovery phenomena. *Radiation Res. Suppl. 6*, 1–29.

Hill, R. F. (1965). Ultraviolet-induced lethality and reversion to prototrophy in *Escherichia coli* strains with normal and reduced dark repair ability. *Photochem. Photobiol. 4*, 563–8.

Hollaender, A., and J. T. Curtis (1935). Effects of sublethal doses of monochromatic ultraviolet radiation on bacteria in liquid suspensions. *Proc. Soc. Exper. Biol. Med. 33*, 61–2.

Hollaender, A., and B. M. Duggar (1938). The effects of sublethal doses of monochromatic ultraviolet radiation on the growth properties of bacteria. *J. Bacteriol. 36*, 17–37.

Howard-Flanders, P. (1975). Repair by genetic recombination in bacteria: Overview. In: *Molecular Mechanisms for Repair of DNA* (P. C. Hanawalt and R. B. Setlow, eds.), Part A. Plenum Press, New York, pp. 265–74.

Jagger, J. (1961). A small and inexpensive ultraviolet dose-rate meter useful in biological experiments. *Radiation Res. 14*, 394–403.

Jagger, J. (1967). *Introduction to Research in Ultraviolet Photobiology*. Prentice-Hall, Englewood Cliffs, N.J., 164 pp.

Kada, T., and H. Marcovich (1963). Sur la siège initial de l'action mutagène des rayons X et des ultraviolets chez *E. coli* K12. *Ann. Inst. Pasteur 105*, 989–1006.

Kelner, A. (1949). Photoreactivation of UV-irradiated *Escherichia coli* with special reference to the dose-reduction principle and to UV-induced mutation. *J. Bacteriol. 58*, 511–22.

Kelner, A. (1953). Growth, respiration, and nucleic acid synthesis in ultraviolet-irradiated and in photoreactivated *Escherichia coli*. *J. Bacteriol. 65*, 252–62.

Kondo, S., and J. Jagger (1966). Action spectra for photoreactivation of mutations to prototrophy in strains of *Escherichia coli* possessing and lacking photoreactivating-enzyme activity. *Photochem. Photobiol. 5*, 189–200.

Krieg, D. R. (1959). A study of gene action of ultraviolet-irradiated bacteriophage T4. *Virology 8*, 80–98.

Lamola, A. A., and T. Yamane (1967). Sensitized photodimerization of thymine in DNA. *Proc. Natl. Acad. Sci. U.S. 58*, 443–46.

Lea, D. E. (1946). *Action of Radiations on Living Cells*. Cambridge University Press, Cambridge, England.

Lederberg, E. M. (1951). Lysogenicity in *E. coli* K-12. *Genetics 36*, 560.

Luria, S. E. (1944). A growth-delaying effect of ultraviolet radiation on bacterial viruses. *Proc. Natl. Acad. Sci. U.S. 30*, 393–6.

Luria, S. E. (1947). Reactivation of irradiated bacteriophage by transfer of self-reproducing units. *Proc. Natl. Acad. Sci. U.S. 33*, 253–64.

Luria, S. E., and R. Latarjet (1947). Ultraviolet irradiation during intracellular growth. *J. Bacteriol. 53*, 149–63.

Lwoff, A., L. Siminovitch, and N. Kjeldgaard (1950). Induction de la production de bactériophages chez une bactérie lysogène. *Ann. Inst. Pasteur 79*, 815–59.

Morowitz, H. J. (1950). Absorption effects in volume irradiation of microorganisms. *Science 111*, 229–30.

Muller, H. J. (1927). Artificial transmutation of the gene. *Science 66*, 84–7.

Patrick, M. H., and R. O. Rahn (1976). Photochemistry of DNA and polynucleotides: Photoproducts. In: *Photochemistry and Photobiology of Nucleic Acids* (S. Y. Wang, ed.), Vol. II. Academic Press, New York, pp. 35–95.

Pettijohn, D., and P. Hanawalt (1964). Evidence for repair replication of ultraviolet-damaged DNA in bacteria. *J. Mol. Biol. 9*, 395–410.

Rahn, R. O., and M. H. Patrick (1976). Photochemistry of DNA; secondary structure, photosensitization, base substitution, and exogenous molecules. In: *Photochemistry and Photobiology of Nucleic Acids* (S. Y. Wang, ed.), Vol. II. Academic Press, New York, pp. 97–145.

Roberts, R. B., and E. Aldous (1949). Recovery from ultraviolet irradiation in *Escherichia coli*. *J. Bacteriol. 57*, 363–75.

Rupert, C. S. (1960). The mechanism of photoreactivation. In: *The Comparative Effects of Radiation* (M. Burton, J. S. Kirby-Smith, and J. L. Magee, eds.). Wiley, New York, pp. 49–61.

Rupert, C. S. (1962). Photoenzymatic repair of ultraviolet damage in DNA. I. Kinetics of the reaction *J. Gen. Physiol. 45*, 703–24. II. Formation of an enzyme-substrate complex. *J. Gen. Physiol. 45*, 725–41.

Rupert, C. S. (1974). Dosimetric concepts in photobiology. *Photochem. Photobiol. 20*, 203–12.

Rupp, W. D., and P. Howard-Flanders (1968). Discontinuities in the DNA synthesized in an excision-defective strain of *Escherichia coli* following ultraviolet irradiation. *J. Mol. Biol. 31*, 291–304.

Sauerbier, W., R. L. Millette, and P. B. Hackett, Jr. (1970). The effect of ultraviolet irradiation on the transcription of T4 DNA. *Biochem. Biophys. Acta 209*, 368–86.

Sedgwick, S. G. (1975). Ultraviolet inducible protein associated with error prone repair in *E. coli* B. *Nature 255*, 349–50.

Setlow, R. B., and W. L. Carrier (1964). The disappearance of thymine dimers from DNA. *Proc. Natl. Acad. Sci. U.S. 51*, 226–31.

Setlow, R. B., and J. K. Setlow (1972). Effects of radiation on polynucleotides. *Ann. Rev. Biophysics Bioengineering 1*, 293–346.

Smith, K. C. (1976). The radiation-induced addition of proteins and other molecules to nucleic acids. In: *Photochemistry and Photobiology of Nucleic Acids* (S. Y. Wang, ed.), Vol. II. Academic Press, New York, pp. 187–218.

Stone, W. S., O. Wyss, and F. Haas (1947). The production of mutations in *Staphylococcus aureus* by irradiation of the substrate. *Proc. Natl. Acad. Sci. U.S. 33*, 59–66.

Sutherland, B. M. (1974). Photoreactivating enzyme from human leukocytes. *Nature 248*, 109–12.

Van Minderhout, L., and J. Grimbergen (1975). A new type of UV-sensitive mutants of phage T4D. *Mutation Res. 29*, 349–62.

Witkin, E. M. (1947). Genetics of resistance to radiation in *Escherichia coli. Genetics 32*, 221–48.

Witkin, E. M. (1956). Time, temperature, and protein synthesis: A study of ultraviolet-induced mutation in bacteria. *Cold Spring Harbor Symp. Quant. Biol. 21*, 123–40.

Witkin, E. M., N. A. Sicurella, and G. M. Bennett (1963). Photoreversibility of induced mutations in a non-photoreactivable strain of *Escherichia coli. Proc. Natl. Acad. Sci. U.S. 50*, 1055–9.

Zimmer, K. G. (1961). *Studies in Quantitative Radiation Biology.* Oliver and Boyd, Edinburgh, 124 pp.

Index